Lucas Vieira Lima
Vinícius A.O. Dittrich

# Ferns and Lycophytes from Lajinha Park, Minas Gerais, Brazil

Lucas Vieira Lima
Vinícius A.O. Dittrich

# Ferns and Lycophytes from Lajinha Park, Minas Gerais, Brazil

## With identification key and brief descriptions

ScienciaScripts

**Imprint**
Any brand names and product names mentioned in this book are subject to trademark, brand or patent protection and are trademarks or registered trademarks of their respective holders. The use of brand names, product names, common names, trade names, product descriptions etc. even without a particular marking in this work is in no way to be construed to mean that such names may be regarded as unrestricted in respect of trademark and brand protection legislation and could thus be used by anyone.

Cover image: www.ingimage.com

This book is a translation from the original published under ISBN 978-3-330-76557-3.

Publisher:
Sciencia Scripts
is a trademark of
Dodo Books Indian Ocean Ltd. and OmniScriptum S.R.L publishing group

120 High Road, East Finchley, London, N2 9ED, United Kingdom
Str. Armeneasca 28/1, office 1, Chisinau MD-2012, Republic of Moldova, Europe
Managing Directors: Ieva Konstantinova, Victoria Ursu
info@omniscriptum.com

Printed at: see last page
**ISBN: 978-620-8-51829-5**

I dedicate this work to the plants.

## ACKNOWLEDGEMENTS

To my advisor and friend Vina, for introducing me to this fantastic world of **pteridology, for the long conversations and advice, for the friendship, for the "vinadas", for the** strengthened **ones** and for having believed in me. Thanks Vina!

To Prof Dr[1] Andrea Pereira Luizi-Ponzo for the monitoring opportunities, the conversations and for adjusting the timetables of the subjects.

To Prof[1] Dr[1] Fátima Regina Gonçalves Salimena for opening the doors of the herbarium for me.

To Prof Dr Alexandre Salino for identifying species with intricate taxonomy.

To all my fellow undergraduates for their company and friendship.

To Flávia Maria Pereira da Costa for her help throughout my degree and for her tips on formatting this work.

To my parents, especially my mother Patrícia Aparecida Vieira Lima, my aunt and uncle Dr José Mauro Teixeira Marinho and Maria das Graças Silva Marinho and my grandmother Márcia Conceição Abras Lima for all their support and investment in my academic life.

To Prof Alan Francisco dos Santos for helping to awaken in me a passion for Biological Sciences.

To Elen, my Little Goose, for her companionship in life and on field trips to carry out this work, for her love, affection and patience. Thank you so much for everything, my love!

To the staff and colleagues at the herbarium, especially over coffee.

To PROPESQ for the grant awarded to carry out this work.

"In the tree by the brook there's a songbird who sings Sometimes all of our thoughts are misgiven...

and it makes me wonder"

Jimmy Page & Robert Plant

## SUMMARY

**Summary**

The general aim of this study was to catalogue the species of pteridophytes that occur in Lajinha Park, Juiz de Fora, Minas Gerais, Brazil. All the species were categorised in terms of life form, a dichotomous key was made to identify the species and brief descriptions of the generic and infrageneric taxa were prepared. Collections were made between October 2013 and May 2014. All the material collected was processed and catalogued in the Leopoldo Krieger Herbarium (CESJ), and available duplicates will be sent to the herbaria of the Federal University of Minas Gerais (BHCB) and the Curitiba City Hall (MBM). A total of 45 species were found in 24 genera, with Thelypteris Schmidel being the most representative, with seven species, followed by Blechnum L., with six, Cyathea J. Sm., with four, Anemia Link, with three, Pteris L., Serpocaulon A.R. Sm., Sticherus C. Presl and Adiantum L., with two species each. The other genera are represented by just one species. Around 90 per cent of the species found are terrestrial or rupicolous and just over 10 per cent are epiphytes, with one species found as an accidental epiphyte. Some exotic species (such as Macrothelypteris torresiana (Gaudich.) Ching, Pteris vitatta L. and Thelypteris dentata (Forssk.) E.P. St John) were found. Identification keys and diagnoses for the genera and species are presented, as well as comments on the environment of occurrence and geographical distribution. This was the first taxonomic study focusing on pteridophytes in the municipality of Juiz de Fora, Minas Gerais.

**Keywords:** monilophytes, floristics, Atlantic Forest, ferns, lycophytes

# CHAPTER 1

## Introduction

Pteridophytes are seedless vascular plants with a heteromorphic life cycle with two distinct phases: the gametophytic phase, which is inconspicuous and ephemeral, and the sporophytic phase, which is larger, more anatomically complex and has a longer life span (Page, 1979). They are not a monophyletic group and therefore should not be treated as a formal taxon. Two large phylogenetically distinct groups of plants are recognised: the Lycophyta and the monilophytes (Pryer et al., 2004).

The Lycophyta are a monophyletic group and are characterised by the presence of microphylls and an adaxial sporangium in the axil of each microphyll. Monilophytes, on the other hand, are characterised by the lateral origin of the root in the endodermis, usually mesarch-like protoxylem in stems and roots, pseudoendospore, plasmodial tapetum and antherozoids with 30 to 1000 flagella (Renzaglia et al. 2000, Schneider et al. 2002).

Pteridophytes occur in the most diverse ecosystems, but there is a high concentration of species in tropical and subtropical mountains (Tryon & Tryon, 1982; Moran, 1995). These plants form an important component of the flora of humid tropical forests, comprising around 10% of the total number of vascular plant species (Grayum & Churchill, 1987). In the epiphytic environment, they always play an important role in the community, with the Polypodiaceae family often appearing as the third richest among vascular epiphytes in Neotropical forests (e.g. Dittrich et al., 1999; Dettke et al., 2008; Bianchi et al., 2012). On the ground, they always surpass any herbaceous angiosperm family in richness (e.g. Foster, 1990; Hammel, 1990; Kozera & Rodrigues, 2005). It is estimated that there are around 17,000 species of pteridophytes in the world (Moran, 2008). According to Moran (1995), there are around 3,000 species of pteridophytes in South America, and Prado (1998) estimates that there are between 1,200 and 1,300 species in Brazil. According to the Brazilian List (Prado & Sylvestre, 2015), 1253 species occur in the country, most of which are found in the Atlantic Forest.

Tropical ecosystems stand out for having the greatest species richness on the planet, even though they only represent around 7% of the world's vegetation cover (tropical forests contain more than half of the world's biota). Within this context is the Atlantic Rainforest, one of the world's biodiversity hotspots, with high species richness and a high level of endemism (Myers et al., 2000).

Minas Gerais has the richest pteridophytic flora in Brazil. According to the Brazilian List, there are 566 species in the state (Prado & Sylvestre, 2015), but Salino & Almeida (2009) have already stated that Minas Gerais has 679 native species, which corresponds to between 52 and 60 per cent of the total species estimated for Brazil.

The municipality of Juiz de Fora, located in the Zona da Mata region of Minas Gerais, is a region of great diversity due to its wide altitudinal range, which is an important factor in increasing the richness of the group in question (Moran, 1995).

There are few exclusive works on the pteridoflora in the Zona da Mata of Minas Gerais (Souza et al., 2012; Souza, 2012), and at least one additional work includes pteridophytes, as well as dealing with other plant groups (Menini et al. 2009). Krieger & Camargo (1990) also contributed to the knowledge of the pteridophytic flora of the Zona da Mata with a list of the region's pteridophytes then deposited in the CESJ herbarium at the Federal University of Juiz de Fora, and Novelino & Oliveira (1999) dealt with the taxonomy of the species of the genus Elaphoglossum Schott ex J. Sm. (Dryopteridaceae) from the Ibitipoca State Park. In the municipality of Juiz de Fora, there is only one published work (Lima & Dittrich 2016) that deals with the species of pteridophytes occurring in 3 forest remnants, but there is no taxonomic treatment for the species and genera Floristic surveys of pteridophytes in Brazil are becoming increasingly frequent (e.g., Salino & Joly (2001), Melo & Salino (2002, 2007), Santiago & Barros (2003), Xavier & Barros (2003), Barros et al. (2004), Santiago et al. (2004), Dittrich et al. (2005), Figueiredo & Salino (2005), Pereira et al. (2007), Schwartsburd & Labiak (2007), Salino & Almeida (2008), Athayde Filho & Felizardo (2010), Gasper & Sevegnani (2010), Matos et al. (2010), Pereira et al. (2011), Forsthofer & Athayde Filho (2012), Mallmann & Schmitt (2014), Barros et al. (2014) ) and Minas Gerais is no exception. However, surveys in urban areas are not very common, even though these areas are closer to research institutions. In the Zona da Mata region of Minas Gerais, there is no published work on the pteridoflora in any urban forest remnant.

Lajinha Park is a public area in the city of Juiz de Fora and, despite this, there have been very few studies of its biota (see Material and methods - characterisation of the study area). It needs to be known in order to conserve it more appropriately, and even the status of the area should change if a Management Plan is established, making it possible for the Park to be included in the SNUC (Juiz de Fora City Council, 2013). From the above, it is clear that the Park still needs many studies to become minimally known from a biological point of view, and this work aims to fill one of these gaps with a complete survey of the pteridoflora of this area, which is the only medium-sized one in the municipality open to public visitation (Prefeitura de Juiz de Fora, 2013). The description of a new species of Anura recently attests to the undeniable biological value of the Park (Santana et al., 2012).

# CHAPTER 2

## Methodology

### 1. Study area

The fieldwork was carried out in Lajinha Park, in the urban area of Juiz de Fora, Minas Gerais.

#### 1.1. Characterisation of the area

The municipality of Juiz de Fora is located in the southeast of the state of Minas Gerais, in the mesoregion known as Zona da Mata Mineira, between the **approximate** geographical coordinates of **21°31'16" and 21°59'59"S and 43°08'50" and 43°41'10"W, and covers an area of** 1,429.88km². The regional climate can be characterised as Koeppen's Cwa type (mesothermal, with hot, rainy summers). The average annual rainfall, measured between 1999 and 2011, is 1627.8mm and the average annual temperature is around 19.3°C for the same period. The altitude varies from 467 to 1104m and the relief is diverse, with concave-convex hills and valleys (Anuário Estatístico de Juiz de Fora, 2012).

Parque da Lajinha was created by municipal decree on 11 July 1978 and renamed Parque da Lajinha by municipal decree in 1982. It has an area of 86.7ha and occupies a region where there used to be a local farm (Fazenda Lajinha). The area was acquired by the municipality through a process of expropriation and has native forest and anthropised areas inside, the forest being of the Montane Seasonal Semideciduous type. In the past, the forest area was more extensive, with the park's woodland then connected to the woods on the **campus of** the Federal University of Juiz de Fora, Morro do Imperador and Fazenda Santa Cândida. It lies to the south-west of the city centre (Juiz de Fora City Hall, 2013), and the entrance to the Park is at the **approximate coordinates of 21°47'34"S and 43°21'58"W. The** altitude varies between approximately 770 and 920 metres inside the park. The area has been little studied to date: the phanerogamic flora has received some attention, but there is still nothing published about it (see Salimena **et al.**, 2005).

**Figura 1** - Satellite image of Lajinha Park, Google Images.

**Sampling to establish floristic composition**

In the area chosen for this study, all the pteridophytes (terrestrial, epiphytic and rupicolous) were collected using the walking method (Filgueiras et al., 1994). For plants out of reach, high pruning shears (loppers) were used. As the forest area is relatively small (ca. 40ha), it was exhaustively explored. The area was visited monthly for eight months from September 2013 to May 2014.

The specimens were collected and prepared according to the usual techniques for pteridophytes (Fidalgo & Bononi, 1984). The exsiccates were deposited in the CESJ herbarium and available duplicates will be sent to the BHCB and MBM herbaria (acronyms according to Thiers, 2013).

The species were determined by the student involved in the project, with the help of specialised literature and, when necessary, with the help of the work's supervisor, who has a doctorate in pteridophyte taxonomy, in addition to comparisons with materials already determined in the CESJ herbarium, which holds the second largest and most important pteridophyte collection in the state. Alexandre Salino (UFMG) helped determine the species of Thelypteris Schmidel, a genus with an intricate taxonomy. This taxonomic study dealt with the groups at a generic level, as the families within the pteridophytes are not very stable and have undergone several changes in their circumscriptions. The descriptions were made on the basis of the materials collected.

# CHAPTER 3

**Results and Discussion**

Forty-five species of pteridophytes were found, distributed in 24 genera, with Thelypteris Schmidel being the most representative, with seven species, followed by Blechnum L., with six, Cyathea J.Sm. with four, Anemia Sw. with three, Pteris L., Serpocaulon A.R. Sm., Sticherus C. Presl and Adiantum L. with two, and the remaining genera with only one species (Table 1). Around 90% of the species found are terrestrial or rupicolous, just over 10% are epiphytes and one species (Pteris schwackeana Christ) was found as an accidental epiphyte.

Exotic species were also collected, such as Macrothelypteris torresiana (Gaudich.) Ching, Pteris vitatta L. and Thelypteris dentata (Forssk.) E.PSt. John, which are characteristic plants of altered environments and can therefore serve as indicators of the park's conservation status.

No rare or threatened species were found. During the collections, dead specimens of Pteridium arachnoideum ((Kaulf.) Maxon were observed in forest edge areas recently shaded due to vegetation growth.

There is a collection record for Pleopeltis minima (Bory) J. Prado & R.Y. Hirai (Krieger s.n., CESJ 2692) in Parque da Lajinha in 1981, but despite collection efforts no other specimens were found, which may serve as an indication of changes in the floristic composition of pteridophytes within this remnant of Atlantic forest.

**Figura 2 A stretch of water inside Lajinha Park (accessed via the Bankers' Club).**

The pollution on the banks of the stretch stands out.

# CHAPTER 4

**Taxonomic treatment**

**Artificial key for identifying the genera of pteridophytes in Lajinha Park,**

**Juiz de Fora, Minas Gerais, Brazil.**

1a. plants with microphylls; one adaxial sporangium in the axil of each microphyll

2a. Isotomously branched plants with indeterminate main axis Phlegmariurus

2b. Plants with anisotomous branching, with a determined main axis

3a Pendulum strobiles ........................................................................................... Straw

3b.Erect strobiles ........................................................................... Pseudolycopodiella

1b. Plants with megaphylls; a few to many sporangia on a leaf 4a.

Apical or oblique annulus, not interrupted by the pedicel

5a. sporangia on a pair of fully fertile basal pinnae ............................................Anaemia

5b.Sporangia in abaxial sori

6a. slender, long-reptant stem; pseudodichotomously branched leaves with dormant axillary buds on the branches

7a.1-bifurcated veins between costa and margin, scales present on the abaxial surface of the blade or at least on the main veins; rhizome and dormant buds with scales Sticherus

7b.2-3-bifurcated veins between costa and margin, abaxial surface of lamina glabrous; rhizome and dormant buds with trichomes

8a Isotomously branched fronds with accessory pinnae at the bifurcations Dicranopteris

8b. Fronds anisotomously branched, absence of accessory pinnae at bifurcations Gleichenella 6b. Stem

robust, arborescent, pinnately branched leaves without buds ...

............................................................................................................... Cyathea

4b. Vertical angle, interrupted by the pedicel

9a.Petiole articulated with the stem, sori without indusium

10a.Single and whole blade

11a.Soros arranged in a row on each side of the coast

12a. Sori rounded, rarely elongated, scales at base of petiole and stem not clathrate Microgramma

12b. Sori oval to elongated, rarely rounded, scales at the base of the petiole and stem clathrate at least in the central portion .................................................................... Pleopeltis

11b. Soros arranged in two or more rows on each side of the ........................................shore

Campyloneurum

10b. Pinatifid or pinnate blade

13a. Free ribs ........ Pecluma

13b. Anastomosing veins

14a. Rhizome long-reptant; sori at the end of a free excurrent vein included in the areoles, clathrate scales ........ Serpocaulon

14b. Rhizome short-reptant; sori at the junction of two excurrent veins included in the areoles; scales not clathrate ........ Phlebodium

9b. Petiole continuous with stem, sori with or without indusium

15a.Marginal or submarginal sori

16a.Rhizome long-reptant, with trichomes; large fronds, usually 1m or more long Hypolepis

16b.Rhizome short-reptant to erect, with scales; fronds small, less than 1m long

17a.Palmately lobed blades ........ Doryopteris

17b. Pinnate to 1-pinnate blades 18a.Ribbed pseudoindusium ........ Adiantum

18b. Pseudoindusium without ribs ........ Pteris

15b. Abaxial sori (sometimes with a very reduced blade of fertile frond)

19a.Elongated sori or sporangia arranged in cenosores

20th Lamina 1-piercing-pinnatifid or more divided

21a.Monomorphic plants with farinaceous excrescence on the abaxial surface Pityrogramma

21b.Dimorphic plants without farinaceous excrescence on the baxial side

22a. Fickle rake ........ Salpichlaena

22b. Non-volatile racket ........ Polybotrya

20b.Lamina 1-pinnate or less divided, rachis not voluble ........ Blechnum

19b.Rounded soros

23a. 2-3-pinned blade ........ Macrothelypteris

23b.Pinnate blade

24a.Scales present on the rachis; petioles with more than two vascular bundles at the base

Ctenitis

24b. No scales on the rachis; petioles with two vascular bundles at the base. Thelypteris

## Adiantum L.

Sp. Pl. 2: 1094. 1753.

Terrestrial or rupicolous plants; stem short or long-reptile or suberect; leaves monomorphic; blade simple to 5-pinnate (or pedunculate), segments flabellate or trapezoidal; petioles glossy, marginal sori, veins free, rarely anastomosing.

The genus differs within the Pteridaceae family in that it is the only one that has sporangia formed on the margin of the recurved leaf blade forming a pseudoindusium and not on the blade beneath it (Lellinger & Prado, 2001). Adiantum is a Pantropical genus with around 200 species, occurring in primary and secondary forests at altitudes ranging from sea level to 5,000 metres in the Andes (Prado et al. 2007). Around 70 species occur in South America (Lellinger & Prado 2001) and 62 in Brazil (Prado & Sylvestre 2015).

**Artificial key for identifying** Adiantum **species in Lajinha Park**

1a. pedunculate blade, glaucous leaflets on the abaxial side, long reptant stem ... A. pentadactylon
1b. Lamina pinnate, olive-green pinnules on the abaxial side, short-reptant stem ..... A. raddianum

Adiantum pentadactylon Langsd. & Fisch., Ic. Fil.: 22, tab. 25. 1810.

Terrestrial or rupicolous plants; long-reptile stem; monomorphic leaves, 70-82 x 3236.7 cm, deltoid blade, peduncled; petiole 35-37.4 cm long, base with scales 0.080.12 cm long, free veins.

Adiantum pentadactylon is a species endemic to Brazil and very common in the southeast, occurring in the states of MG, RJ, SP, PR, SC and RS. It can be found near riverbanks or in ravines in the forest, and is characterised as heliophilous or shade-tolerant (Winter et al, 2011).

Material examined: BRAZIL. Minas Gerais, Juiz de Fora, Lajinha Park, water trail, 21°47'28.07"S 43°22'08.26"W, 815 m, 29.X.2013, L.V.Lima 15(CESJ).

Adiantum raddianum C. Presl., Tent. Pterid.:158. 1836.

Terrestrial plants; short-reptile stem; monomorphic leaves, 15.2-31.5 x 2.3-15.6 cm, deltoid lamina, 2-3-pinnate; petiole 7.3-13.7 cm long, base with scales 0.25-0.3 cm long; free veins.

A. raddianum is recognisable by its flabellate pinnules and reniform sori. According to Winter et al. (2011), it occurs from Mexico to Uruguay and is a widely cultivated and distributed species, occurring practically in all Brazilian states. It is a plant that is indifferent to light conditions, shade-tolerant and grows in the most diverse habitats, including anthropised environments.

Material examined: BRAZIL. Minas Gerais, Juiz de Fora, Lajinha Park, water trail, 21°47'28.07"S 43°22'8.26"W, 815m, 29.X.2013, L.V.Lima 16 (CESJ).

Anaemia (L.) Sw.

Syn. Phil. 6, 155. 1806.

Terrestrial or rupicolous plants; stem short-reptile, horizontal or ascending; leaves haemidymorphic to dimorphic; blade 1-3-pinnate; sori usually arranged in a pair of fertile pinnae; veins free or anastomosing.

According to Mickel & Smith (2004), the genus Anemia includes approximately 120 species, most of which occur in tropical America, ten in Africa and one in India. According to these authors, the genus has its centre of diversity in Brazil (around 70 species). The genus is widespread in Brazil, occurring in almost all states, with the exception of Rio Grande do Norte (Mickel et al., 2015)

**Artificial key for identifying** Anemia **species in Lajinha Park**

1a.Flattened petiole

2a. anastomosing veins .......................................................................................... A. phyllitidis

2b.Free ribs ........................................................................................................... A villosa

1b. Cylindrical petiole .......................................................................................... A. hirsuta

Anemia hirsuta (L.) Sw., Syn. Fil. (Swartz) 156. 1806.

Terrestrial or rupicolous plants; stem suberect to erect; leaves 12-23.5 x 1.8-4.6 cm, blade oblong-straight; petiole 3.4-4.8 cm long, base with trichomes, veins free.

A. hirsuta is a species with great phenotypic plasticity, with small to large fronds. This species is capable of hybridisation, including with A. phylitidis, also found in Lajinha Park. It is distributed throughout most of tropical America from low altitudes to 4000m (Mickel & Smith 2004). In Brazil, it is distributed throughout the Southeast and partially in the Northeast and Centre-West regions (Mickel et al., 2015).

Material examined: BRAZIL. Minas Gerais, Juiz de Fora, Lajinha Park, waterfall trail, 21°47'37.93"S 43°22'9.37"W, 777 m, 30.I.2014, L.V.Lima 29 (CESJ).

Anemia phyllitidis (L.) Sw., Syn. Fil. 155. 1806

Terrestrial plants; decumbent stem; leaves 30-42.4 x 11-12.5 cm, deltoid blade, pinnate; petiole 17-127 cm long, base with trichomes; veins partially anastomosing.

A. phyllitidis is a species with a wide distribution in the Neotropics and is easily distinguished by its anastomosing venation, 1-pinnate blade and conforming apical pinna, making it a morphologically variable species. In Brazil, it is distributed in almost all states except for a few in the North and Northeast regions (Mickel et al. 2015).

Material examined: BRAZIL. Minas Gerais, Juiz de Fora, Lajinha Park, water trail, 21°47'28.07"S 43°22'8.26"W, 815m, 05.X.2013, L.V.Lima 02 (CESJ).

Anemia villosa Humb. & Bonpl. ex Willd., Sp. Pl. 5: 92. 1810

Terrestrial plants; decumbent stem; leaves 11.2-15.3 x 4-8 cm, narrow-triangular blade, 2-pinnate-pinnatifid to pinnate-pinnatifid; petiole 5.6-11.5 cm long, base with trichomes; free veins.

A. villosa is a species with a limited distribution in South America (Santos & Sylvestre, 2006), and in Brazil it is distributed in all the states of the Southeast, AC, AM, RO, TO, AL, BA, CE, PE, SE, DF, GO, MS, MT.
Material examined: BRAZIL. Minas Gerais, Juiz de Fora, Lajinha Park, water trail, 21°47'28.07"S 43°22'8.26"O, 815 m, 29.X.2013, L.V.Lima 17 (CESJ).

Blechnum L.

Sp. Pl. 2: 1077. 1753.

Terrestrial or rupicolous plants, rarely hemiepiphytes or holoepiphytes; short to long-ribbed stems, decumbent or erect, or even arborescent; monomorphic or dimorphic leaves, pinnatisect to 1-pinnate blade, petiole always present, adaxially furrowed and continuous with the stem; abaxial sori, elongated along the main veins, with indusium; free to partially anastomosing veins.

The genus Blechnum has ca. 150-200 species, of which approximately 50 occur in Tropical America (Tryon & Tryon, 1982). Around 30 species occur in Brazil, eight of which are endemic (Dittrich & Salino, 2015). Molecular studies have shown that the genus is clearly paraphyletic or polyphyletic. In Lajinha Park it is represented by six species.

**Artificial key for identifying** Blechnum **species in Lajinha Park**

1a.Dimorphic leaves

2a. Blades partially or completely attached to the rachis ............................................ B. acutum

2b. Leaves, at least the basal ones, petiolate or sessile, not adnate to the root. B. cordatum 1b. Monomorphic leaves

3a.Arborescent plants with linear, nigrescent petiole base scales 2.5 cm or more long B. brasiliense

3b.Non-arborescent plants; petiole base scales triangular to narrowly triangular, brown, up to 1 cm long

4a. Base of the leaf blade gradually reduced with the basal leaflets reduced to auricles B. ..........................................................................................................polypodioides

4b. Base of leaf lamina truncated; basal leaflets not or only slightly reduced

5a. first pair of pinnae with proximal acroscopic side partially or totally adnate to the rachis B. austrobrasilianum

5b. First pair of pinnae with completely free proximal acroscopic side ................................ ........................................................................................................................ B. occidentale

Blechnum acutum (Desv.) Mett., Ann. Sci. Nat. Bot., ser. 5, 2: 225.1864

Hemiepiphytic plants; long-reptile stem; dimorphic leaves, fertile 46-75 x 5.7- 18.3 cm; sterile 44.4-102 x 12.2-27 cm; fertile blade pinnate, elliptic, sterile pinnatisect or 1-pinnate at the base, lanceolate; petiole of fertile leaves 14.5 cm long; of sterile leaves 5.1-23.0 cm long, base with scales 10-13 mm long; free veins.

This species is widely distributed in the Neotropical region. In Brazil it occurs from Rio Grande do Sul to Bahia. Among the species in the area, it is mainly characterised by its hemiepiphytic habit, long-reptant stem with bicoloured scales and dimorphic fronds.

Material examined: BRAZIL. Minas Gerais, Juiz de Fora, Parque da Lajinha, trilha do clube dos bancários, 21°47'5,92"S, 43°22'8,76"O, 869m, 8.VI.2014, L.V.Lima 57 (CESJ).

Blechnum austobrasilianum de la Sota, Bol. Soc. Arg. Bot. 16(3): 248. 1975

Terrestrial plants, erect to decumbent stem, monomorphic leaves, 35-54 x 5-7 cm; linear-lanceolate blade, pinnate in the basal portion, pinnatisect in the middle and apical portions, truncate base; petiole 12-22 cm long, base with scales 0.15-0.25 cm long; free veins.

According to Dittrich (2005), this species is distributed in South America in Bolivia, north-eastern and north-western Argentina and south-eastern and southern Brazil, and together with B. occidentale and B. polypodioides is the subject of much confusion because they are close species and hybridise with each other.

B. austrobrasilianum can be distinguished from B. polypodioides by having the base of the blade truncated or with reduced leaflets, but never lobed, while B. polypodioides has the base of the blade strongly reduced with the last leaflets reduced to auricles. B. austrobrasilianum differs from B. occidentale in having the acroscopic side of the basal pinna adnate to the rachis, while in B. occidentale it is free.

Material examined: BRAZIL. Minas Gerais, Juiz de Fora, Parque da Lajinha, water trail, 21°47'28.07"S 43°22'8.26"O, 815m, 29.X.2013, L.V.Lima 13 (CESJ); idem, waterfall trail, 21°47'37.93"S 43°22'9.37"O, 777m, 30.I.2014, L.V.Lima 27 (CESJ).

Blechnum brasiliense Desv., Ges. Naturf. Freunde Berlin Mag. Neuesten Entdeck. Gesammten Naturk. 5: 330. 1811

Terrestrial plants, arborescent stem, monomorphic leaves, 167-173 x 38-39 cm; blade oblanceolate, pinnate at the base and pinnatisect in the median and apical portions, base gradually reduced; petiole 10- 18.5 cm long, base with scales 2.5 -3.0 cm long, free veins.

B. brasiliense differs from other species of the genus in Brazil in that it is the only arborescent, monomorphic species with nigrescent scales at the base of the petiole. This species is widely distributed in the Neotropics, where it generally grows in degraded areas, such as along trails, roads and riverbanks. It is a very common species in Brazil and occurs from sea level to 1800 metres above sea level (Dittrich 2005).

Material examined: BRAZIL. Minas Gerais, Juiz de Fora, Lajinha Park, water trail, 21°47'28.07"S 43°22'8.26"O, 815m, 21.XI.2013, L.V.Lima 23 (CESJ).

Blechnum cordatum (Desv.) Hieron., Hedwigia 47: 239. 1908

Terrestrial plants; erect stem; dimorphic leaves, fertile 30-127 x 8-10 cm, sterile 18-105 x 13-16 cm; fertile blade ovate-lanceolate, pinnate, base truncate; sterile blade lanceolate, pinnate, base truncate, base of petiole with scales 0.6-1.5 cm long, veins free.

Material examined: BRAZIL. Minas Gerais, Juiz de Fora, Parque da Lajinha, water trail, 21°47'28.07"S 43°22'8.26"O, 815 m, 21.XI.2013, L.V. Lima 21 (CESJ), idem, water trail, 21°47'28.07"S 43°22'8.26"O,

815m, 05.X.2013, L.V. Lima 09 (CESJ).

Blechnum occidentale L., Sp. Pl. 1077. 1753

Terrestrial plants; erect stem; monomorphic leaves, 42-45 x 9-11 cm; deltoid blade, pinnate in the basal and median portions, pinnatisect in the apical portion, truncate base; base of the petiole with scales 0.2-0.75 cm long; free veins.

Widely distributed in the Neotropics, in Brazil it occurs in all the states of the South and Southeast regions, GO, MS, MT, BA, CE, PE, RO.

Material examined: BRAZIL. Minas Gerais, Juiz de Fora, Parque da Lajinha, water trail, 21°47'28.07"S 43°22'8.26"O, 815 m, 05.X.2013, L.V.Lima 03 (CESJ); idem, waterfall trail, 21°47'37.93"S 43°22'9.37"O, 777m, 30.I.2014, L.V.Lima 33 (CESJ).

Blechnum polypodioides Raddi, Sci. Bol. 3: 294. 1819

Terrestrial plants; erect stem; monomorphic leaves, 15-35 x 2.5-5.5 cm; linear-lanceolate blade, predominantly pinatissecta, reduced base; petiole base with scales 0.1-0.3 cm long; free veins.

According to Dittrich (2005), this species is very common in Brazil, especially in the South and Southeast regions, and is distributed throughout most of tropical America

Material examined: BRAZIL. Minas Gerais, Juiz de Fora, Parque da Lajinha, water trail, 21°47'28.07"S 43°22'8.26"O, 815 m, 05.X.2013, L.V. Lima 5 (CESJ); idem, waterfall trail, 21°47'37.93"S 43°22'9.37"O, 777m, 30.I.2014, L.V. Lima 25 (CESJ).

Campyloneurum C. Presl

Suppl. Tent. Pterid. 189. 1836

Epiphytic plants, occasionally terrestrial or rupicolous; short to long-reptant stem; simple blade; abaxial sori, rounded, without indusium; anastomosing veins.

According to Léon (1993), Campyloneurum is defined by its areolate venation, with costal areoles with an included excurrent vein and entire or divided non-costal areoles. The genus is characterised by species with an entire or 1-pinnate blade. According to Schneider et al. (2004), the genus is part of a monophyletic group that also includes the genus Niphidium, from which it can be differentiated by having two rows of sori between adjacent lateral veins, versus one row in Niphidium.

Campyloneurum acrocarpon Fée, Crypt. Vasc. Brésil 1:115, t. 35, f. 3. 1869.

Terrestrial plants; long-reptile stem; monomorphic leaves, 45-55 x 5-6 cm; lanceolate blade, simple, entire, base attenuated; petiole base with scales 0.15-0.1 cm long; anastomosing veins.

C. acrocarpon can be easily recognised by its long petiole (15-20 cm) and the presence of rhizome-like scales on the abaxial side of the costa (Vasques & Prado, 2011). This species is endemic to Brazil and is distributed throughout the states of the South and Southeast regions, while in the Northeast it only occurs in the state of Bahia.

Material examined: BRAZIL. Minas Gerais, Juiz de Fora, Parque da Lajinha, trilha do clube dos bancários, 21°47'53,10"S 43°22'9,30"O, 853m, 8.VI.2014, L.V.Lima 49 (CESJ)

Ctenitis (C. Chr.) C. Chr.

In: Verdoorn, Man. Pterid. 544.1938

Terrestrial plants; stem erect or suberect, rarely short-reptant; blade 1-pinnate-pinnatifid; sori abaxial, rounded, with or without indusium; free veins.

Ctenitis is a pantropical genus with around 150 species, whose centre of diversity is southern Mexico and Guatemala (Mickel & Smith 2004). The closest genus to Ctenitis is Megalastrum, which differs in that it has whitish, strigose trichomes on the adaxial surface of the axes and hydathodes on the abaxial side of the blade. However, in recent molecular analyses (e.g. Schuettpelz & Pryer, 2007), the genus Ctenitis appears relatively isolated, forming a monophyletic group within the lineage of eupolipodioid ferns.

Ctenitis distans (Brack.) Ching, Bull. Mem. Inst. Biol. Bot. 8: 277. 1938

Terrestrial plants; erect stem; monomorphic leaves, 132.8-133.7 x 26.4-27.6 cm; ovate-lanceolate blade, truncate base; base of petiole with scales 0.58-0.78 cm long; free veins.

This species is endemic to Brazil and occurs in the South, Southeast and Northeast regions of the country (Salino & Almeida, 2015). It is characterised by its generally patent proximal pinnae, dark brown to black rachis scales, relatively rigid with a jagged margin and filiform apex, inframedian sori and conspicuous indusium present.

Material examined: BRAZIL. Minas Gerais, Juiz de Fora, Parque da Lajinha, trilha do clube dos bancários, 21°47'51,40"S 43°22'9,65"O, 833m, 08.VI.2014, L.V.Lima 42 (CESJ)

Cyathea J. Sm.

Acad. Roy. Sci. (Turin) 5: 416. 1793.

Terrestrial plants; stipe-like erect stem; 1-pinnatifid to 2-pinnatifid blade; petiole inert, spinescent or muricate, with marginalised scales at the base; sori abaxial, rounded, with or without indusium; free veins.

Cyathea is a genus with around 115 species (Mickel & Smith, 2004). In Brazil there are around 39 species and two varieties, 13 of which are endemic (Windisch & Santiago, 2015)

**Artificial key for identifying** Cyathea **species in Lajinha Park**

1a.2-pinned blade ........................................................................................ C. corcovadensis

1b.2-pinnatifid lamina

2a.Sori with globose indusium ........................................................................ C. delgadii

2b. Soros without indusium

3a. Abaxial surface indumentum with two types of trichomes (hirsute and villous), presence 3b. Indumentum on the abaxial side with only one type of trichome, absence of spines on the rachis. C. phalerata

Cyathea corcovadensis (Raddi) Domin, Pterid. Isl. Dominica 262. 1929

Terrestrial plants; erect stipe-like stem; monomorphic leaves, 3.20 x 1.37 m; oblong blade, 2-pinnate, truncate base; base of petiole with scales 2.2-2.58 cm long; free veins.

This species is endemic to Brazil and is distributed throughout the South, Southeast and Northeast regions of the country. It is a species that shows great variation in the cut of its pinnules, which range from entire to slightly pinnatifid (Fernandes, 1997).

Material examined: BRAZIL. Minas Gerais, Juiz de Fora, Parque da Lajinha, trilha do clube dos bancários, 21°47'29.65"S 43°22'55.84"O, 881m, 12.IV.2014, L.V.Lima 53 (CESJ)

Cyathea delgadii Sternb., Vers. Fl. Vorwelt 1: 47. t. B. 1820.

Terrestrial plants; erect stipe-like stem; monomorphic leaves, 2.95-3.52 x 1.18-1.24 m; blade ovate, base truncate; base of petiole with scales 1.5-2 cm long; free veins.

C. delgadii is widely distributed throughout Tropical America, occurring from Costa Rica to Argentina. In Brazil, it is distributed in all regions. According to Fernandes (2003) this species is one of the most frequent arborescent ferns in southern Brazil, where it occurs in practically all types of vegetation, from sea level to 1500 metres above sea level. In Lajinha Park, this species is widespread in its domain, occurring in a wide variety of habitats, both on forest edges and inland, in shaded or open areas, closer or not to stretches of water.

Material examined: BRAZIL. Minas Gerais, Juiz de Fora, Parque da Lajinha, water trail, 21°47'28.07"S 43°22'8.26"O, 815m, 05.X.2013, L.V.Lima 01 (CESJ)

Cyathea microdonta (Desv.) Domin, Pteridophyta 263. 1929.

Terrestrial plants; erect stipe-like stem; monomorphic leaves, 2.58-2.97 x 1.08-1.17 m; blade ovate, base truncate; base of petiole with scales 1.4-2 cm long; free veins.

It is a species with a wide distribution throughout tropical America. In Brazil, it is distributed in all of the country's states, with the exception of DF, MA, PR, RS and RN (Windisch & Santiago, 2015).

Material examined: BRAZIL. Minas Gerais, Juiz de Fora, Parque da Lajinha, trilha do clube dos bancários, 21°47'29,65"S, 43°22'55,84"O, 881m, 12.IV.2014, L.V.Lima 54 (CESJ) Cyathea phalerata Mart., Denkschr. Bot. Ges. Regensb. 2: 146, t. 2, f. 3. 1822.

Terrestrial plants; erect stipe-like stem; monomorphic leaves, 2.65-2.89 x 0.98-1.18 m; blade ovate, base truncate; base of petiole with scales 0.6-2 cm long; free veins.

This species is endemic to Brazil and is distributed in all the states of the South and Southeast regions, DF, GO, MT, Al, BA, CE, PE, AC. It occurs preferentially in the interior of humid forests or near streams in drier forests, from sea level to 1500 metres above sea level (Fernandes 2003). In Lajinha Park it was found near a stretch of water and in an adjacent wetland.

Material examined: BRAZIL. Minas Gerais, Juiz de Fora, Parque da Lajinha, trilha do clube dos bancários, 21°47'29.65"S 43°22'55.84"O, 881m, 08.VI.2014, L.V.Lima 60 (CESJ)

Dicranopteris Bernh.

Neues J. Bot. 1(2): 38. 1806 [1805].

Terrestrial plants; long-reptile stem; monomorphic leaves; pseudodichotomously branched, isotomous blade with a pair of accessory leaflets at each fork; dormant buds covered in trichomes, abaxial sori without indusium; free veins.

Dicranopteris is a pantropical genus with around 10 species (Mickel & Smith, 2004) characterised by the presence of accessory spines opposite the branches, bifurcated to trifurcated veins, sori with 15-16 sporangia and rhizomes with trichomes.

Dicranopteris flexuosa (Schrad.) Underw. Bull. Torrey Bot. Club 34: 254. 1907.

Terrestrial plants; long-reptile stems; leaflets 10-28.5 x 2-5 cm, linear; free veins.

This species is distributed throughout most of Brazil and has not yet been recorded only in Acre and Piauí.

Material examined: BRAZIL. Minas Gerais, Juiz de Fora, Lajinha Park, water trail, 21°47'28.07"S 43°22'8.26"O, 815m, 22.III.2014, L.V. Lima & E.J.O. Furtado 34 (CESJ).

Doryopteris J. Sm.

J. Bot. (Hooker) 4: 162. 1841

Terrestrial or rupicolous plants; compact, erect stem; dimorphic fronds; marginal sori, elongated, with pseudoindusium; anastomosing or free veins.

According to Mickel & Smith (2004), the genus has around 30 species, most of which are found in the eastern belt of Brazil. The genus is characterised by its palmate blade and marginal sori.

Doryopteris nobilis (T. Moore) C. Chr., Ind. Fil. 244. 1905.

Terrestrial plants; erect stem; dimorphic leaves, sterile fronds 13.4-27.3 x 6.7-8.6 cm; peduncle blade with buds at the base, 3-lobed; fertile fronds 31.4-35 x 10.3-11.8 cm; buds at the base; peduncle blade, 5-lobed; base of petiole with scales 0.39-0.46 cm long; anastomosing veins.

This species is distributed throughout the states of the South and Southeast, MS, BA. It can be distinguished by the buds at the apex of the petiole.

Material examined: BRAZIL. Minas Gerais, Juiz de Fora, Parque da Lajinha, trilha da água, 21°47'28.07"S 43°22'8.26"O, 815 m, 22.III.2014, L.V.Lima & E.J.O.Furtado 38 (CESJ); idem, 21°47'28.07"S 43°22'8.26"O, 815m, 21.XI.2013, L.V.Lima & E.J.O.Furtado 22 (CESJ).

Gleichenella Ching Sunyatsenia

5: 276. 1940.

Gleichenella is a monospecific genus with a wide distribution in Brazil and the Neotropics (Mickel & Simth, 2004).

Gleichenella pectinata (Willd.) Ching, Sunyatsenia 5: 276.1940

Terrestrial plants; long-reptile stem; monomorphic fronds, pseudodichotomously branched, anisotomous; pinnules 7.5-11 x 0.9-1.4 cm, clinear; dormant bud covered in trichomes; free veins.

G. pectinata differs from Dicranopteris flexuosa by the absence of accessory pinnae at the bifurcations and by an anisotomous branching pattern.

Material examined: BRAZIL. Minas Gerais, Juiz de Fora, Parque da Lajinha, trilha do clube dos bancários, 21°47'52.80"S 43°22'9.93"O, 861m, 12.IV.2014, L.V.Lima 45 (CESJ)

Hypolepis Bernh.

Neues J. Bot. 1(2): 34. 1806 [1805].

Terrestrial plants; long-reptile stem; 2-4-pinnate blade; marginal sori, rounded to oblong, usually with a small indusium; free veins.

Hypolepis is made up of around 50 species and is distributed in the Neotropics and temperate southern regions (Moran, 1995b)

Hypolepis repens (L.) C. Presl, Suppl. Tent. Pterid. 162. 1836.

Terrestrial plants; long-reptile stem; monomorphic leaves, 154.2-163.4 x 30-31 cm; deltoid blade, 3-pinnate-pinnatifid, truncate base; petiole, base of petiole with trichomes; free veins.

H. repens is characterised by its spinescent axes and reddish-brown petiole. It is distributed throughout tropical America, in Brazil it occurs in RS, SC, ES, MG, RJ, SP, AL, CE, PE (Assis, 2008).

Material examined: BRAZIL. Minas Gerais, Juiz de Fora, Parque da Lajinha, trilha do clube dos bancários, 21°47'29,65"S 43°22'55,84"O, 881m, 12.IV.2014, L.V.Lima 41 (CESJ).

## Macrothelypteris (H. Itô) Ching

Acta Phytotax. Sin. 8: 308. 1963.

Terrestrial plants; stem short-reptile or suberect; blade 2-pinnate to 2-pinnatifid; sori abaxial, rounded, usually with a small indusium; free veins.

Macrothelypteris comprises around nine species in the Old World and has one species that is widely naturalised and spread around the world. The genus can be easily differentiated from Thelypteris because has a more jagged blade (Mickel & Smith, 2004).

Macrothelypteris torresiana (Gaudich.) Ching, Acta Phytotax. Sin. 8: 310. 1963.

Terrestrial plants; erect stem; monomorphic leaves, 49.4-98 x 18-92.2 cm; deltoid blade, 2-pinnate-pinnatifid, truncate base; base of petiole with scales 5.9-7.7 mm long; free veins.

This species can be easily distinguished from the other species in the park by having two vascular bundles at the base of the petiole, a 2-pinnate or more divided blade, a longer or only slightly reduced proximal pinna, but never vestigial, and the presence of acicular hyaline trichomes on the abaxial side of the blade.
It is a widespread and naturalised species in the Neotropics. In Brazil it occurs in PR, RS, SC, ES, MG, RJ, SP, DF, GO, MS, MT, AL, BA, CE, PA, PE, RN.

Material examined: BRAZIL. Minas Gerais, Juiz de Fora, Parque da Lajinha, trilha da cachoeira,

21°47'37,93"S 43°22'9,37"O, 777m, 30.I.2014, L.V.Lima 30 (CESJ); idem, 21°47'37.93"S 43°22'9.37"O, 777m, 30.I.2014, L.V.Lima 31 (CESJ); idem, banking club trail, 21°47'29.65"S 43°22'55.84"O, 881m, 12.IV.2014, L.V.Lima 40 (CESJ).

Microgramma C. Presl

Suppl. Tent. Pterid. 213, t. 9, f. 7. 1836.

Epiphytic plants, occasionally terrestrial or rupicolous; long-reptant stem; simple fronds; abaxial sori, rounded to elongated, without indusium; anastomosing veins.

Microgramma is a genus comprising around 30 species distributed in the Neotropics and around 2 species in Africa and the Indian Ocean islands (Salino et al., 2008).

Microgramma squamulosa (Kaulf.) de la Sota, Opera Lilloana 5: 59. 1960.

Epiphytic plants; long-reptant stem; dimorphic leaves, sterile fronds 8-12.4 x 1-1.2 cm, fertile fronds 12.5-13.6 x 0.8-0.9 cm; linear-lanceolate blade, simple, entire, base arising; base of petiole with scales 4.3-5 mm long; anastomosing veins.

This species is distributed throughout South America (Brazil, Peru, Bolivia, Paraguay, Argentina and Uruguay (Tryon & Stolze, 1993). In Brazil it occurs in all the states of the South and Southeast regions, in the Northeast only in the state of Bahia. Despite showing great morphological variation, it can be recognised by the densely ciliated scales at the apex of the rhizome, with long cilia, and by the presence of lanceolate scales on the coast of the sterile fronds (Almeida, 2014).

Material examined: BRAZIL. Minas Gerais, Juiz de Fora, Parque da Lajinha, trilha do clube dos bancários, 21°47'29.65"S 43°22'55.84"O, 881m, 12.IV.2014, L.V.Lima 48 (CESJ)

Vasc. straw & Franco

Bol. Soc. Broter. II, 41: 24. 1967

Terrestrial plants; erect, upright stem with indeterminately growing main branches; apical strobiles on lateral branches, sessile, recurved to pendulous.

Palhinhaea is a genus with a wide distribution mainly in tropical and subtropical regions, with pioneer species from disturbed places growing especially in tropical mountains (Arana & Ollgaard, 2012).

Palhinhaea cernua (L.) Vasc. & Franco, Bol. Soc. Broter. 2, 41: 25. 1967.

Terrestrial plants; erect, upright stem 76-82cm x 0.3-0.4, with obvious main branch; helically arranged trophophylls, alternately verticillate sporophylls, 0.4-0.6cm long, 0.4-9mm wide. Sporophylls smaller than trophophylls.

This species has a Pantropical distribution. In Brazil, it is distributed in all regions of the country (Windisch et al., 2015). It differs from P. camporum (a closely related species) having softer, pendulous secondary aerial axes, more widely spaced microphylls and 4-12 mm strobili, while P. camporum has stiffer secondary aerial axes and more densely grouped microphylls and 14-24 mm strobili (Arana & Ollgaard, 2012).

Material examined: BRAZIL. Minas Gerais, Juiz de Fora, Lajinha Park, waterfall trail, 21°47'37.93"S 43°22'9.37"O, 777m, 05.X.2013, L.V.Lima 04 (CESJ)

Pecluma M.G.Price

Amer. Fern J. 73: 109. 1983.

Epiphytic plants, less commonly rupicolous or terrestrial; stem short to long-reptate; blade pinnatisecta with linear pinnules; sori abaxial, rounded, without indusium; ribs free.

Pecluma is an exclusively Neotropical genus and comprises around 30 species (Price, 1983). The genus, according to Mickel & Smith (2004), seems to be closer to Polypodium species without scales on the lamina, particularly with the free venation group such as P. hartwegianum, P. chiapense and P. diplotrichum.

Pecluma robusta (Fée) M. Kessler & A.R. Sm., Candollea 60 (1): 281. 2005.

Terrestrial or epiphytic plants; long-reptile stem; monomorphic leaves, 27-165 cm x 07-14 cm; oblong blade, base with reduced pinnules; base of petiole with scales 3 x 5-8 mm; free veins.

A species found from Central America to Bolivia, Brazil, northern Argentina and Paraguay (Labiak, 2005). In Brazil it occurs in all the states of the South and Southeast regions, in the Northeast it is present only in the state of Bahia. Pecluma robusta is characterised by the presence of sparse trichomes on the abaxial side of the blade, except around the sori with the formation of "crowns" of trichomes (Rolim & Salino, 2008).

Material examined: BRAZIL. Minas Gerais, Juiz de Fora, Parque da Lajinha, trilha do clube dos bancários, 21°47'58.40"S 43°22'9.06"O, 868m, 08.VI.2014, L.V.Lima 44 (CESJ); idem, 21°47'58.40"S 43°22'9.06"O, 868m, 12.IV.2014, L.V.Lima 52 (CESJ).

Phlebodium (R.Br.) J.Sm.

J. Bot. (Hooker) 4: 58. 1841

Epiphytic plants, less commonly rupicolous; stem short to long-reptant; lamina pinnatifid to pinnatissecta; sori abaxial, rounded, without indusium; veins anastomosing.

Phlebodium is a genus that occurs exclusively in the Neotropics and comprises around 4 species (Mickel & Smith, 2004).

Phlebodium pseudoaureum (Cav.) Lellinger, Amer. Fern. J. 77: 101. 1987.

Epiphytic plants; long-reptile stem; monomorphic leaves, 19.5-44 x 7.5-17 cm; deltoid blade, pinatissecta, truncate base; petiole base with scales 5.3-7.6 cm long; anastomosing veins.

This species is widespread in Brazil, with records for all states. It is characterised by its essentially glabrous petiole, lamina, costa and costulae, and by having only one row of sori between the costa and the margin of the pina.

Material examined: BRAZIL. Minas Gerais, Juiz de Fora, Parque da Lajinha, trilha do clube dos bancários, 21°47'59.70"S 43°22'8.87"O, 861m, 08.VI.2014, L.V.Lima 58 (CESJ)

Phlegmariurus (Herter) Holub

Preslia 36 (1): 17, 21. 1964

Terrestrial or epiphytic plants; erect or pendulous stem, branched or rarely simple; monomorphic or subdimorphic microphylls with very reduced sporophylls arranged in terminal branches.

Phlegmariurus is a genus with around 300 to 400 species distributed in tropical and temperate zones of the planet, with the highest level of richness in the Andes (0llgaard, 1995).

*Phlegmariurus reflexus* (Lam.) B. 0llg., Rodriguesia 63: 481. 2012.

Terrestrial plants; stem prostrate at the base and erect in the apical portion, 4.5-16.5 cm x 0.10.2 mm;

isodichotomous branching; microphylls arranged spirally throughout the plant, 57 x 0.5-1 mm; sporophylls similar to trophophylls.

This species is distributed throughout tropical America from Mexico to Argentina. In Brazil, it is distributed in all the states of the South and Southeast regions, DF, MS, AM, PA. It is a common species in disturbed areas and in Lajinha Park it was found in the open, on ravines.

Material examined: BRAZIL. Minas Gerais, Juiz de Fora, Parque da Lajinha, trilha do clube dos bancários, 21°47'29.65"S, 43°22'55.84"O, 881 m, 05.X.2013, L.V.Lima 08 (CESJ)

Pityrogramma Link

Handbuch 3: 19. 1833

Terrestrial plants; erect stem; monomorphic leaves; 1-3-pinnate blade, abaxial side usually covered with a thin layer of yellowish or white material, elongated sori along the veins, the latter free.

Pityrogramma is a genus with around 16 predominantly neotropical species that grow from sea level to altitudes of 1000 - 2000 m, sometimes up to 3500 m (Tryon & Tryon, 1982). According to Mickel & Smith (2004), they are plants of disturbed places such as roadsides.

Pityrogramma calomelanos (L.) Link, Handbuch 3: 20. 1833.

Terrestrial plants; erect stem; monomorphic leaves, 7-32 x 6-8.5 cm; deltoid blade, pinnate-pinnatisect, truncate base; petiole base with scales 0.3-0.5 cm long; free veins.

A species with a Neotropical distribution. In Brazil it is distributed in all states (Prado, 2015). It is characterised by a thin layer of yellowish or white material on the abaxial side of the leaf blade and by the sori distributed in rows along the veins, as well as the glabrous blade.

Material examined: BRAZIL. Minas Gerais, Juiz de Fora, Parque da Lajinha, trilha da cachoeira, 21°47'37,93"S 43°22'9,37"O, 777m, 29.X.2010, L.V.Lima 11 (CESJ).

Pleopeltis Humb. & Bonpl. ex Willd.

Sp. Pl., ed. 4, 5(1): 211. 1810

Epiphytic plants, less commonly rupicolous; stem long-reptile; lamina simple, entire to pinnatifid, rarely pinnatifid; sori abaxial, rounded to oblong, without indusium; veins anastomosing, rarely free.

Pleopeltis is a genus comprising around 90 species with a predominantly neotropical distribution. It is characterised by the presence of round peltate scales on the lamina and sori and venation with areolae with free vein included (Mickel & Smith, 2004).

**Artificial key for identifying** Pleopeltis **species in Lajinha Park**

1a. simple and entire lamina, oblong sori ..................................................................... P. astrolepis

1b. Blade pinatissecta, round sori ............................................................................. P. hirsutissima

Pleopeltis astrolepis (Liebm.) E.Fourn., Mexic. Pl. 1: 87.1872

Epiphytic plants; long-reptant stem; monomorphic leaves, 4.5-12.7 x 0.62-0.87 cm; blade linear- lanceolate, simple, entire, base arising; petiole base with scales 0.30.5 mm long; veins immersed.

This species is distributed throughout tropical America. In Brazil, it is distributed throughout the states of the South and Southeast, GO, AL, BA, CE, PA, PE (Labiak & Hirai, 2015a). It is a very common species in Lajinha Park, where it has been found in various forophytes. It is distinguished by its elongated sori, short petiole and small stem scales, smaller than 0.5 mm.

Material examined: BRAZIL. Minas Gerais, Juiz de Fora, Lajinha Park, waterfall trail, 21°47'37.93"S 43°22'9.37"O, 777m, 22.III.2014, L.V.Lima & E.J.O Furtado 35 (CESJ)

Pleopeltis hirsutissima (Raddi) de la Sota, Darwiniana 45: 239. 2007

Epiphytic plants; short-reptile stem; monomorphic leaves, 8-12.5 x 0.8-1.1 cm; blade ovate, base truncate; base of petiole with scales 0.3-0.8 cm long; veins immersed.

This species is distributed in tropical America. In Brazil, it is distributed throughout the states of the South, Southeast and Centre-West regions (Labiak & Hirai, 2015).

Material examined: BRAZIL. Minas Gerais, Juiz de Fora, Parque da Lajinha, trilha do clube dos bancários, 21°47'5,67"S 43°22'8,95"O, 876m, 08.VI.2014, L.V.Lima 59 (CESJ)

## Polybotrya Humb. & Bonpl. ex Willd.

Sp. Pl., ed. 4, 5(1): 99. 1810

Terrestrial or hemiepiphytic plants; long-reptile stem; dimorphic fronds; sporangia in cenosores, without indusium; anastomosing or free veins.

Polybotrya is a genus with around 35 neotropical species, and is distinguished by having strongly dimorphic fronds in which the fertile fronds look like skeletons of the sterile (photosynthesising) fronds, and a rhizome covered in scales (Moran, 1987).

Polybotrya speciosa Schott, Genera Filicum tab.7.1834. 244. 1905

Hemiepiphytic plants; long-reptant stem; dimorphic leaves, sterile fronds 30-120 x 26-80 cm; blade lanceolate to ovate, 2-3-pinnate; base of petiole with scales 8-20 x 0.51.5 mm; anastomosing veins.

Note: Despite the collection efforts, no fertile specimens were found, so only the sterile frond was described in this work.

This species is endemic to the Serra do Mar mountains and differs from other Polybotrya species in that the rhizome scales are strongly denticulated and red and the multicellular and branched paraphyses are patent, concoloured and smooth; the adaxial surface is subglabrous with trichomes only in the furrows, the abaxial surface is subglabrous, sparsely to moderately hairy only on the veins (0. 1-0.4 mm and 2-4 cells).1-0.4 mm and 2-4 cells), and clearly branched paraphyses (Canestraro, 2014).

Material examined: BRAZIL. Minas Gerais, Juiz de Fora, Parque da Lajinha, trilha do clube dos bancários, 21°47'29.65"S 43°22'55.84"O, 881m, 12.IV.2014, L.V.Lima 46 (CESJ)

## Pseudolycopodiella Holub

Folia Geobot. Phytotax. 18: 441. 1983

Terrestrial plants; stem with prostrate main branches with indeterminate growth, radicant, with horizontal lateral branches and erect dorsal branches; apical strobili on dorsal branches, isolated strobili conspicuously differentiated from the peduncle.

Pseudolycopodiella is a genus with around 10 species distributed in temperate and tropical regions, with the largest number of species in South America (Arana & Ollgaard, 2012).

Pseudolycopodiella caroliniana (L.) Holub Folia Geobot. Phytotax. 18: 441. 1983

Terrestrial plants; branched stem with prostrate main branches with indeterminate growth, with erect branches 16-18.4 wide x 0.1-0.2 in diameter; microphylls of the erect branch more acute than those of the main branch, 0.3-0.7 x 0.1-0.3 cm. Sporophylls smaller than trophophylls.

This species is widely distributed in Brazil (Windisch et al, 2015). According to Mickel & Smith (2004), it is distinguished by its flattened terrestrial stems and sparse microphylls on the peduncle.

Material examined: BRAZIL. Minas Gerais, Juiz de Fora, Lajinha Park, waterfall trail, 21°47'37.93"S 43°22'9.37"O, 777m, 30.I.2014, L.V.Lima & E.J.O Furtado 37 (CESJ)

Pteris L.

Sp. Pl. 2: 1073. 1753

Terrestrial plants, less commonly rupicolous; stem erect to short-reptant; blade pinnate or more jagged; marginal sori, elongated, with false indusium; veins anastomosing or free.

Pteris is a Pantropical genus with around 200 species, characterised by sori in a marginal commissure connecting the ends of the veins, with filamentous paraphyses, a strongly differentiated indusium with a recurved margin and tetrahedral or globose, trilete spores.

**Artificial key for identifying** Pteris **species in Lajinha Park**

1a.Free venation ........................................................................................................ P. vittata

1b.Anastomosing venation ........................................................................... P. schwackeana

Pteris vittata L., Sp. Pl. 2: 1074. 1753.

Terrestrial plants; stem erect; leaves monomorphic, 21.2-47.6 x 5.8-11.5cm; blade elliptic-ovate, pinnate, base truncate; base of petiole with scales 3.5-4.2mm long; veins free.

A naturalised species in the Neotropics, with records from the United States to Argentina. It has a wide distribution in Brazil. It is a species characteristic of altered environments, found in open, well-lit places, occurring from sea level to around 2000 metres above sea level. It is characterised by having a 1-pinnate blade, linear leaflets with a truncate base and free venation (Prado & Windisch, 2000).

Material examined: BRAZIL. Minas Gerais, Juiz de Fora, Lajinha Park, waterfall trail, 21°47'37.93"S 43°22'9.37"O, 777m, 30.I.2014, L.V.Lima 26 (CESJ)

Pteris schwackeana Christ in Schwacke, Pl. Nov. Min.:26.1900

This species is endemic to Brazil and is distributed throughout the Southeast (ES, MG, RJ, SP) and Northeast (PE) regions. It is characterised by the blade1-pinnate, with oblong to linear, entire pinnules.

The venation is areolate, with large areoles near the costa, smaller ones near the margin and all the veins free above the last areoles on the margin. The indumentum consists of dark brown to red scales on the petiole and part of the rachis, and scabrous surfaces on the petiole and rachis. These last two characteristics differentiate it from P. splendens, a species closely related to P. schwackeana (Prado & Windisch, 2000).

Note: Despite collection efforts, only one young individual of P. schwackeana was found, which is why this paper does not include a description of this species.

Salpichlaena J.Sm., J. Bot. (Hooker) 4: 168. 1841

Terrestrial plants; long-reptile stem; 2-pinnate blade; abaxial cenosorium, adjacent and parallel to the main veins; indusium present; free veins.

Salpichlaena is a Neotropical genus with two species: S. volubilis and S. hookeriana (Kuntze) Alston. The genus differs from the others in Blechnaceae in that it has voluble leaves (Giudice et al., 2008).

Salpichlaena volubilis (Kaulf.) J.Sm. J. Bot. (Hooker) 4: 168. 1841

Epiphytic plants; long-reptile stem; dimorphic leaves, ca. 15 m long, fertile pinnules 1135 x 0.5-3 cm, sterile pinnules 12-18 x 1.5-2.5 cm; blade with truncated base; free veins.

This species is widely distributed in tropical America (Giudice et al., 2008). In Brazil, it is distributed throughout all regions (Dittrich & Salino, 2015). It is characterised by its scandent habit, paleaceous and cartilaginous pinnule margins and linear sori along both sides of the coast (Giudice et al., 2008).

Material examined: BRAZIL. Minas Gerais, Juiz de Fora, Parque da Lajinha, trilha do clube dos bancários, 21°47'59,30"S 43°22'8,63"O, 859m, 22.III.2014, L.V.Lima & E.J.O Furtado 61(CESJ).

Serpocaulon A.R.Sm., Taxon 55 (4): 924-927.2006

Epiphytic or rupicolous plants, infrequently terrestrial; stem long-reptant, rarely short-reptant; blade entire or pinnatifid to pinnate; sori abaxial, rounded, without indusium; veins anastomosing.

Serpocaulon is a genus with around 40 neotropical species, with the greatest diversity in the humid forests of the Andes and south-eastern Brazil. The genus is monophyletic and is characterised by regular anastomosing venation, long-reptant rhizomes with clathrate scales, usually peltate, and sori without paraphyses (Smith et al., 2006).

**Identification key for** Serpocaulon **species in Lajinha Park**

1st pinnate blade ............................................................................ S. fraxinifolium

2b. Blade pinatissecta ......................................................................................... S. catharinae

Serpocaulon catharinae (Langsd. & Fisch.) A.R.Sm., Taxon 55: 928. 2006.

Epiphytic plants; long-reptile stem; monomorphic leaves, 23-56.5 x 11.4-14.3 cm; narrow-triangular blade, pinnatisect, truncate base; base of petiole with scales 0.260.28 cm long; anastomosing veins.

This species is distributed in part of South America (Argentina, Brazil, Paraguay and Uruguay) (Labiak & Prado, 2008). In Brazil, it is distributed throughout the states of the South, Southeast and Northeast regions (Al, BA, PA, PE) (Labiak & Hirai, 2015b). It is distinguished by its pinatissecta blade, wrinkled rhizome containing whitish wax deposits and dark brown scales sparsely distributed on the rhizome (Labiak & Prado, 2008).

Material examined: BRAZIL. Minas Gerais, Juiz de Fora, Parque da Lajinha, trilha do clube dos bancários, 21°47'29,65"S 43°22'55,84"O, 881m, 12.IV.2014, L.V.Lima 43 (CESJ).

Serpocaulon fraxinifolium (Jacq.) A.R.Sm., Taxon 55: 928. 2006.

Epiphytic plants; long-reptile stem; monomorphic leaves, 23-56.5 x 62.3-79.4 cm; narrow-triangular blade, pinnate, truncate base; base of petiole with scales 0.31-0.33 cm long; anastomosing veins.

This species is widely distributed in the Neotropics, where it grows preferentially in the interior of humid forests at relatively low altitudes. In Brazil, it is distributed throughout the South (PR, SC), Southeast (ES, MG, RJ, SP), Centre-West (DF, GO), Northeast (BA, PE) and North (RO) regions (Labiak & Hirai, 2015b). It is distinguished by its 1-pinnate blade, long-reptant rhizome, sparsely coated with lanceolate scales, and sori in 3-4 (5) series between the costa and margin (Labiak & Prado, 2008).

Material examined: BRAZIL. Minas Gerais, Juiz de Fora, Parque da Lajinha, trilha do clube dos bancários, 21°47'29,65"S 43°22'55,84"O, 881m, 12.IV.2014, L.V.Lima 50 (CESJ).

Sticherus C. Presl

Suppl. Tent. Pterid. 51. 1836.

Terrestrial plants; long-reptile stem; monomorphic leaves; pseudodichotomously branched blade, isotomic; dormant buds covered in scales, abaxial sori without indusium; free veins.

Sticherus is a Pantropical genus with around 95 species, characterised by a rhizome and buds with scales, generally bifurcated segment venules, 2-5(-7) sporangia per sorus, pseudodichotomously branched blades with the last segments pinnatifid to pinnate and monolete spores (Gonzales & Kessler, 2011).

**Identification key for** Sticherus **species in Lajinha Park**

1st blade 1 to 2 times forked ...................................................................... Sticherus nigropaleaceus

2b. Blade 3 or more times forked .................................................................. Sticherus lanuginosus

Sticherus lanuginosus (Fée) Nakai, Bull. Natl. Sci. Mus. 29: 20. 1950.

Terrestrial plants; long-reclining stem; leaflets 10.4-14.5 x 2.8-3.4 cm, linear; blade 3 or more times forked, with aphelia on dormant buds; free veins.

This species is distributed throughout tropical America. In Brazil, it is distributed in all the states of the South and Southeast regions, in the Northeast it is only present in the state of Bahia. S. lanuginosus is a common species, which for a long time was known under the name S. penniger, but Gonzales & Kessler (2011) in a review of the genus and study of the type, found that S. penniger is synonymous with S. pruinosus. It is characterised by its thick rhizome (3.5-6.5 mm in diameter), which is generally glabrous, and by its erect fronds with short, ascending branches, showing a remarkably narrow angle between the first and second (third) order axes. In addition, the blade is 4-5 forked and the scales present on the buds and axes of the plant are delicate, hyaline and long fimbriated (Gonzales, 2003).

Material examined: BRAZIL. Minas Gerais, Juiz de Fora, Lajinha Park, water trail, 21°47'28.07"S 43°22'8.26"O, 815m, 05.X.2013, L.V.Lima 06 (CESJ).

Sticherus nigropaleaceus (J.W. Sturm) Prado & Lellinger, Amer. Fern J. 86: 98. 1996

Terrestrial plants; long-reptile stems; 28-39 x 11-12.6 cm linear leaflets; blades 12 times forked; no aphelia covering the dormant buds; free veins.

A common species in humid forests between 300 and 800 metres above sea level in south-eastern Brazil, with isolated populations in Amazonas and Mato Grosso do Sul, as well as in La Paz (2500-3750 metres above sea level) (Gonzales & Kessler, 2011). It can be distinguished by the dark brown to nigrescent scales on the buds and rachis, on the abaxial side.

Material examined: BRAZIL. Minas Gerais, Juiz de Fora, Lajinha Park, water trail, 21°47'28.07"S 43°22'8.26"O, 815m, 29.X.2013, L.V.Lima 20 (CESJ).

Thelypteris Schmidel

Icon. Pl. (ed. Keller) 45, t. 11, 13. Oct 1763

Terrestrial plants, rarely rupicolous or epiphytic; stem short to long-reptant, erect or still erect; blade pinnate or more commonly pinnate-pinnatifid, base of petiole in transverse section with two vascular bundles at the base; sori abaxial; veins free or anastomosing.

The genus Thelypteris is subcosmopolitan, with around 875 species, of which approximately 300 occur in the Neotropical region (Salino & Semir, 2004a). Five subgenera occur in Brazil: Thelypteris subg. Amauropelta (Kunze) A.R. Sm., T. subg. Cyclosorus (Link) C.V. Morton, T. subg. Goniopteris (C. Presl) Duek, T. subg. Meniscium (Schreber) C.F. Reed and T. subg. Steiropteris (C. Chr.) Iwats. (Salino & Semir, 2004a). Four subgenera occur in Lajinha Park.

**Artificial key for identifying** Thelypteris **species from Lajinha Park**

1a. pinnate blade .......................................................................................................... T. serrata

1b. Pinnate blade

2a. Gradually reduced blade base

3a. Long-reptant stem ........................................................................ T. rivularioides

3b. Erect to erect stem

4a. uncinate trichomes present on the petiole, and/or between the abaxial surface of the segments between the veins

5a. stem scales with septiform trichomes ............................................. T. regnelliana

5b Stem scales with uncinate trichomes ................................................ T. amambayensis

4b. Uncinate trichomes absent on the petiole, and/or between the abaxial surface of the segments between the veins ....................................................................................... T. opposita

2b. Truncated blade base

6a.Basal veins of adjacent segments joining directly at the inlet ..............................................
............................................................................................................................ T. mexiae

6b. Basal veins of adjacent segments joining an excurrent vein towards the inlet ... T. dentata

Thelypteris amambayensis (Christ) Ponce, Candollea 55:310. 2000.

Terrestrial plants, erect stem, monomorphic leaves 55-83 x 12.5-14.5 cm; linear-lanceolate blade, reduced base; base of petiole with scales 0.25-0.30 cm long; free veins.

This species is only distributed in Argentina, Paraguay and Brazil. In Brazil, it is distributed throughout the states of the South and Southeast regions. T. amambayensis is related to T. regnelliana and can be distinguished by having uncinate trichomes on the scales, petiole and rachis, while T. regnelliana has uncinate trichomes on the abaxial side of the back, ostulae and laminar tissue. In addition, T. amambayensis has buds at the base of the distal pinnules (Salino & Semir, 2004b).

Material examined: BRAZIL. Minas Gerais, Juiz de Fora, Parque da Lajinha, water trail, 21°47'28.07"S 43°22'8.26"O, 815m, 05.X.2013, L.V.Lima 10 (CESJ); idem, waterfall trail, 21°47'37.93"S 43°22'9.37"O, 777m, 30.I.2014, L.V.Lima 28 (CESJ).

Thelypteris dentata (Forssk.) E.P. St. John, Amer. Fern J. 26:44. 1936

Terrestrial plants, short-reptile stem; monomorphic leaves 82.5-87 x 18-22.6 cm; blade lanceolate to elliptical, base truncate; petiole 34.4-48.5 cm long, base with scales 0.460.50 cm long; free veins.

It is an introduced and widely naturalised species in the New World, currently occurring from the south of the USA to Argentina, tropical and subtropical Africa, Asia and the Pacific Islands. In Brazil, it occurs in all the states of the South and Southeast regions, AL, BA, CE, MA, PA, PE, DF, GO, MS, MT (Salino & Semir, 2002). According to Salino and Semir (2002) the oldest collection from Brazil examined in this study dates from 1927, which is from the state of São Paulo (Brade 7710). However, Smith (1971) cites an older collection from 1908 from the state of Pará (Hernandez s.n.).

Material examined: BRAZIL. Minas Gerais, Juiz de Fora, Parque da Lajinha, water trail, 21°47'28.07"S 43°22'8.26"O, 815m, 29.X.2013, L.V.Lima 12 (CESJ); idem, waterfall trail, 21°47'37.93"S 43°22'9.37"O, 777m, 30.I.2014, L.V.Lima 32 (CESJ).

Thelypteris mexiae (C. Chr. ex Copel.) Ching, Bull. Fan Mem. Inst. Biol., Bot. 10: 252. 1941

Terrestrial plants; erect stem; monomorphic leaves, 98-125 x 28-30.8 cm; lanceolate blade, base truncated; base of petiole with scales 0.3-0.4 cm long; free veins.

Endemic to Brazil, known only from the states of Pernambuco, Minas Gerais and Espírito Santo (Salino & Almeida, 2015).

Material examined: BRAZIL. Minas Gerais, Juiz de Fora, Parque da Lajinha, trilha do clube dos bancários, 21°47'29.65"S 43°22'55.84"O, 881 m, 12.IV.2014, L.V.Lima 51 (CESJ)

Thelypteris opposita (Vahl) Ching, Bull. Fan Mem. Inst. Biol., Bot. 10: 251. 1941

Terrestrial plants; erect stem; monomorphic leaves; 86-108 x 10-12.5 cm; elliptical blade, reduced base; base of petiole with scales; free veins.

Species with a Neotropical distribution (Greater and Lesser Antilles, Costa Rica, Panama, Venezuela, Colombia to Paraguay). In Brazil, it occurs in all the states of the South, Southeast and Centre West regions, BA, AC, RO. Thelypteris opposita is characterised by the gradually reduced base of the blade, with 10-20 pairs of reduced leaflets, often reaching close to the stem, making the petiole very short, reaching up to 2 cm in length (Salino & Semir, 2004b).

Material examined: BRAZIL. Minas Gerais, Juiz de Fora, Lajinha Park, water trail, 21°47'28.07"S

43°22'8.26"O, 815m, 05.X.2014, L.V.Lima 07 (CESJ).

Thelypteris regnelliana (C.Chr.) Ponce, Darwiniana 33: 264. 1995

Terrestrial plants; stem decumbent to erect; leaves monomorphic, 86-115 x 19-21.2 cm; lamina elliptic, base reduced; base of petiole with scales; veins free.

This species is restricted to the south-east and south of Brazil, occurring in all the states of these regions (Salino & Semir, 2004b).

Material examined: BRAZIL. Minas Gerais, Juiz de Fora, Parque da Lajinha, water trail, 21°47'28.07"S 43°22'8.26"O, 815m, 29.X.2013, L.V.Lima 19 (CESJ), idem, 21°47'28.07"S 43°22'8.26"O, 815m, 22.III.2014, L.V.Lima & E.J.O.Furtado 36 (CESJ).

Thelypteris rivularioides (Fée) Abbiatti, Rev. Mus. La Plata, Secc. Bot. 9: 19. 1958.

Terrestrial plants; long-reptile stem; monomorphic leaves, 33-109 x 12-14 cm; elliptical blade with gradually reduced base; base with scales, free veins.

Species with a distribution restricted to southern South America. In Brazil it occurs in all the states of the South and Southeast regions, BA, DF, GO. This species occurs preferentially in sunny places, with very humid soil, often waterlogged, and on the edges of riparian and paludosa forests (Salino & Semir, 2004b). In Lajinha Park it was found at the edge of the forest.

Material examined: BRAZIL. Minas Gerais, Juiz de Fora, Lajinha Park, water trail, 21°47'28.07"S 43°22'8.26"O, 815m, 21.XI.2013, L.V.Lima 24 (CESJ).

Thelypteris serrata (Cav.) Alston, Kew Bull. 1932: 309. 1932

Terrestrial plants; long-reptile stem; monomorphic leaves 30-140 cm x 25-30 cm; pinnate blade, lanceolate, truncate base; petiole base with scales; anastomosing veins.

This species ranges from the United States to Argentina. In Brazil it occurs in all states except Rio Grande do Sul. Thelypteris serrata can be distinguished from other species occurring in Brazil by the serrate to uncinate-serrate margins on the upper third at least (Salino & Semir, 2004).

Material examined: BRAZIL. Minas Gerais, Juiz de Fora, Lajinha Park, water trail, 21°47'28.07"S 43°22'8.26"O, 815 m, 29.X.2013, L.V.Lima 14 (CESJ).

# CHAPTER 5

## Conclusion

The vegetation suffers from edge effects, especially in the areas where it comes into contact with public roads, and the park is constantly subjected to anthropogenic actions such as weeding, material disposal and rubbish thrown into watercourses. The conservation conditions are deplorable, especially in the areas that are not open to visitors. A large amount of rubbish and waste material was found in running watercourses. All in all, the number of species recorded for this small remnant of Atlantic forest, situated in the middle of a city with more than half a million inhabitants, is very significant.

The survey of pteridophytes carried out in this remnant of Atlantic Rainforest provided a better understanding of the current state of the municipal flora and could help to characterise the forest formation of this remnant, which has suffered constant anthropogenic disturbances, indirectly interfering with its composition, thus providing data for further studies in Lajinha Park and for studies of other remnants in Juiz de Fora.

## Bibliographical references


ALMEIDA, T.E. **Systematic and phylogenetic studies on the genus** Microgramma **C. Presl (Polypodiaceae - Polypodiopsida)** PhD Thesis, Universidade Federal de Minas Gerais, Belo Horizonte, 2014.

AMABRÓSIO, S. T. & BARROS, I. C. L.. **Pteridophytes from a remnant area of Atlantic Forest in the state of Pernambuco, Brazil.** Acta Botanica Brasilica 11(2): 105-113, 1997.

Statistical Yearbook of Juiz de Fora. 2012. http://www.cps.ufjf.br/anuarios/Anuario2012/index.html Accessed on 19.03.2012, 18h45min.

ARANA, M. D.; 0LLGAARD, B. **Revisión de las Lycopodiaceae: Embryopsida, Lycopodiidae) de Argentina y Uruguay.** Darwiniana, nueva serie, v. 50, n. 2, p. 266-295, 2012.

ASSIS, F. C. **The Family Dennstaedtiaceae Pic. Serm. sensu lato (Polypodiopsida) in the state of Minas Gerais, Brazil.** Master's dissertation. Institute of Biological Sciences, Federal University of Minas Gerais, Belo Horizonte. 84 p. 2008.

ATHAYDE FILHO, F.P. & FELIZARDO, M.P.P. **Floristic and ecological analysis of the ferns and lycophytes of the main spring of the Pindaíba River, Mato Grosso.** Research, Botany 61: 229-244, 2010.

BARROS, I.C.L., XAVIER, S.R.S., PIETROBOM, M.R., LOPES, M.S., LUNA, C.P.L., SOUZA, G. S. & SANTIAGO, A.C.P. **Pteridophytic flora of the São João da Várzea Ecological Reserve (Recife - Pernambuco - Brazil).** Boletim da Sociedade Broteriana 71: 187-194, 2004.

BARROS, I. C. L.; CANTARELLI, L. C.; FARIAS, R. P.; PEREIRA, A. F. N.; DA SILVA, I. A. A.. Vertical distribution of epiphytic ferns in an Atlantic Forest fragment in the north-east of Brazil. IHERINGIA,

Sér. Bot, 69(1), 143-153, 2014.

BIANCHI, J.S.; BENTO, C. M.; KERSTEN, R. A. **Vascular epiphytes of an ocotone area between the Dense and Mixed Ombrophilous Forests, in the Marumbi State Park, PR ponto** Estud. Biol., Ambiente Divers.;34(82):37-44, 2012.

CANESTRARO, B. K. **The genus** Polybotrya **(Dryopteridaceae) in the Atlantic Forest of Brazil.** Master's thesis, Federal University of Paraná, Paraná. 2014

DETTKE, G. A.; ORFRINI, A. C.; MILANEZE-GUTIERRE, M.A.M. **Floristic composition and distribution of vascular epiphytes in an altered remnant of Semideciduous Seasonal Forest in Paraná, Brazil.** Rodriguésia 59: 859-872, 2008.

DITTRICH, V.A.O.. **Taxonomic studies on the genus Blechnum in the south-east and south of Brazil.** Doctoral thesis, Universidade Estadual Paulista, Rio Claro, 2005.

DITTRICH, V.A.O.; SALINO, A. **Blechnaceae in Lista de Espécies da Flora do Brasil.** Rio de Janeiro Botanical Garden. Available at: <http://floradobrasil.jbrj.gov.br/jabot/floradobrasil/FB90785>. Accessed on: 31 Mar. 2015.

DITTRICH, V.A.O., KOZERA, C. & SILVA, S.M. **Floristic survey of the vascular epiphytes of Barigui Park, Curitiba, Paraná, Brazil.** Iheringia, Sér. Bot. 52: 11-21, 1999.

DITTRICH, V.A.O., WAECHTER, J.L. & SALINO, A. **Species richness of pteridophytes in a montane Atlantic rain forest plot of Southern Brazil.** Acta Botanica Brasilica 19(3): 519525, 2005.

FERNANDES, I. **Taxonomy and phytogeography of Cyatheaceae and Dicksoniaceae in the South and Southeast regions of Brazil.** PhD Thesis, University of São Paulo, 1997.

FERNANDES, I. **Taxonomy of the representatives of Cyatheaceae from northeastern Brazil.** Research, Botany 53: 7-53, 2003.

FIGUEIREDO, J.B. & SALINO, A. **Pteridophytes from four private natural heritage reserves in the south of the metropolitan region of Belo Horizonte, Minas Gerais, Brazil.** Lundiana 6(2): 83-94, 2005.

FILGUEIRAS, T.S., BROCHADO, A.L., NOGUEIRA, P.E. & GUALLA II, G.F. **Walking - an expeditious method for qualitative floristic surveys.** Cadernos de Geociências 12: 39-43, 1994.

FORSTHOFER, M. & ATHAYDE FILHO, F.P. **Floristics and ecological aspects of ferns and lycophytes along the Cachoeirinha stream, Nova Xavantina-MT**. Research, Botany 63: 149-164, 2012.

FOSTER, R. B. **The floristic composition of the Rio Manu floodplain forest.** In: A. H. Gentry (ed.). Four neotropical rainforests. pp. 99-111. Yale University Press, New Haven, 1990.

FOSTER, R. B. & HUBBEL, S. T. **The floristic composition of the Barro Colorado Island forest.** In: A. H. Gentry (ed.). Four neotropical rainforests. pp. 85-98. Yale University Press, New Haven, 1990.

GALINDO-LEAL, C. & CÂMARA, I.G. **Status of the Atlantic Forest Hotspot: a synthesis**. In: Galindo-Leal, C. & Câmara, I.G. (Eds.). Atlantic Forest: Biodiversity, threats and perspectives. pp. 3-11. SOS

Foundation Mata Atlântica, São Paulo, 2005.

GASPER, A.L. & SEVEGNANI, L. **Lycophyta and ferns of the Serra do Itajaí National Park, Vale do Itajaí, SC, Brazil.** Hoehnea 37(4): 755-767, 2010.

GIUDICE, G. E., LUNA, M. L., CARRIÓN, C., & DE LA SOTA, E. R. **Revision of the genus Salpichlaena J. Sm. (Blechnaceae, Pteridophyta).** American Fern Journal, v. 98, n. 2, p. 49-60, 2008.

GONZALES, M.J. **A taxonomic revision of the genus *Sticherus* (Gleicheniaceae - Pteridophyta) in the Neotropics.** Thesis (Doctorate in Maths and Science), University of Gottingen, Gottingen, Germany, 2003.

GONZALES, J., KESSLER, M. **A synopsis of the Neotropical species of** Sticherus **(Gleicheniaceae), with descriptions of nine new species.** Phytotaxa, v. 31, p. 1-54, 2011.

GRAYUM, M. H. & CHURCHILL, H. W. **An introduction to the pteridophyte flora of Finca La Selva, Costa Rica.** American Fern Journal 77: 73-89, 1987.

HAMMEL, B. **The distribution of diversity among families, genera and habit types in the La Selva Flora.** In: A. H. Gentry (ed.). Four neotropical rainforests. pp. 75-84. Yale University Press, New Haven, 1990.

KOZERA, C. & RODRIGUES, R.R. **Submontane dense ombrophilous forest: floristics and structure of the lower stratum.** In: Marques, M.C.M. & Britez, R.M. (Eds.). Natural history and conservation of Ilha do Mel. Ed. UFPR, Curitiba. Pp. 103-123, 2005.

KRIEGER, L. & CAMARGO, R.F.N. **Pteridophytes from the Zona da Mata of Minas Gerais found in the Herbarium of the Federal University of Juiz de Fora.** Proceedings of the 36th National Botanical Congress, Curitiba. p. 287-306, 1990.

LABIAK, P.H. 2005. Polypodiaceae. In Flora of the Federal District, Brazil (T.B. Cavalcanti & E.E. Ramos, eds). Embrapa-Cenargem, Brasília. v.4, p.159-181.

LABIAK, P.H.; HIRAI, R.Y. Pleopeltis **in Lista de Espécies da Flora do Brasil.** Rio de Janeiro Botanical Garden. Available at: <http://floradobrasil.jbrj.gov.br/jabot/floradobrasil/FB91723>. Accessed on: 01 Apr. 2015a

LABIAK, P.H.; HIRAI, R.Y. Serpocaulon in Lista de Espécies da Flora do Brasil. Rio de Janeiro Botanical Garden. Available at: <http://floradobrasil.jbrj.gov.br/jabot/floradobrasil/FB91749>. Accessed on: 01 Apr. 2015b

LABIAK, P.H., PRADO, J. **New combinations in Serpocaulon and a provisional key for the Atlantic Rain Forest species.** American Fern Journal, v. 98, n. 3, p. 139-159, 2008.

LELLINGER, D.B. & PRADO, J. **The group of** Adiantum gracile **in Brazil and environs.** American Fern Journal 91: 1-8, 2001.

LEÓN, B. **A taxonomic revision of the fern genus Campyloneurum (Polypodiaceae).** Doctoral thesis, 1993.

LIMA, L.V. & DITTRICH, V.A.O. **Lycophytes and monilophytes from three remnant areas of Semideciduous Seasonal Forest in the municipality of Juiz de Fora, Minas Gerais, Brazil.** Research, Botany n° 69:183-198,

2016.

LIMA, R. A. F., DITTRICH, V.A.O., SOUZA, V.C., SALINO, A., BREIER, T.B. & AAGUIAR, O.T. **Vascular flora of Carlos Botelho State Park, São Paulo, Brazil.** Biota Neotropica 11: 173-214, 2011.

List of Brazilian Flora Species. 2013. (Updated constantly.) http://floradobrasil.jbrj.gov.br/. Accessed on 06.03.2013, 15:48.

MALLMANN, I. T.; SCHMITT, J. L. **Richness and floristic composition of the fern community in the riparian forest of the Cadeia River, Rio Grande do Sul, Brazil.** Ciência Florestal 24.1 97-109, 2014.

MATOS, F.B., AAMORIM, A.M. & LABIAK, P.H. **The ferns and lycophytes of a montane tropical forest in southern Bahia, Brazil.** Journal of the Botanical Research Institute of Texas 4(1): 333-346, 2010.

MELO, L.C.N. & SALINO, A. **Pteridophytes from two forest areas of the Rio Doce basin in the state of Minas Gerais, Brazil.** Lundiana 3(2): 129-139, 2002.

MELO, L.C.N. & SALINO, A. **Pteridophytes in forest fragments of APA Fernão Dias, Minas Gerais, Brazil.** Rodriguésia 58(1): 207-220, 2007.

MENINI NETO, L., MATOZINHOS, C.N., ABREU, N.L., VALENTE, A.S.M., ANTUNES, K., SOUZA, F.S., VIANA, P.L. & SALIMENA, F.R.G. **Non-arboreal vascular flora of a grotto forest in Serra da Mantiqueira, Zona da Mata de Minas Gerais, Brazil.** Biota Neotropica 9(4): 1-13, 2009.

MICKEL, J.; BARROS, I.C.L.; SANTIAGO, A.C.P.; PEREIRA, A.F. de N.; LABIAK, P.H. **Anemiaceae in Lista de Espécies da Flora do Brasil.** Rio de Janeiro Botanical Garden. Available at: <http://floradobrasil.jbrj.gov.br/jabot/floradobrasil/FB90589>. Accessed on: 31 Mar. 2015.

MICKEL, J.; SMITH, A.R. **The Pteridophytes of Mexico**. Memoirs of the New York Botanical Garden, New York, v.88, p. 1-1055, 2004

MORAN, R. C. **Monograph of the neotropical fern genus Polybotrya (Dryopteridaceae).** Illinois Nat. Hist. Surv. Bull. 34: 1-138, 1987.

MORAN, R.C. **The importance of mountains to pteridophytes, with emphasis on Neotropical montane forests.** In: Churchill, S.P., Balslev, H., Forero, E., Luteyn, J.L. (Eds.). pp. 359363. Biodiversity and conservation of Neotropical montane forests. New York Botanical Garden, New York, 1995.

MORAN, R. C.. Hypolepis Bernh. In: Moran, R.C. & Riba, R. (eds.). Mesoamerican Flora. Vol. 1. Psilotaceae to Salvineaceae. Universidad Autónoma de México, Ciudad de México. Pp 153156. 1995b.

MYERS, N., MITTERMEIER, R.A., MITTERMEIER, C.G., FONSECA, G.A.B. & KENT, J. **Biodiversity hotspots for conservation priorities**. Nature 403: 853-858, 2000.

NOVELINO, R.F. & OLIVEIRA, J.E.Z. **Flora of Ibitipoca State Park, Minas Gerais, Brazil: Elaphoglossaceae (Pteridophyta).** Brazilian Institute of the Environment and Renewable Natural Resources, Brasília, 1999.

0LLGAARD, B. **Diversity of Huperzia (Lycopodiaceae) in Neotropical Montane Forests,** In S. P. Churchill (ed.), Biodiversity and Conservation of Neotropical Montane Forests, pp. 349358. New York: New York

Botanical Garden Press. 1995.

PAGE, C.N. **The diversity of ferns: an ecological perspective.** In: Dyer, A.F. (Ed.). The experimental biology of ferns. Academic Press, London. Pp. 10-56, 1979.

PEREIRA, A.F.N., BARROS, I.C.L., XAVIER, S.R.S. & SANTIAGO, A.C.P. **Floristic composition and ecology of the pteridoflora of Atlantic Forest fragments (Gurjaú Ecological Reserve, Cabo de Santo Agostinho, Pernambuco, Brazil).** Brazilian Journal of Biosciences 5: 489-491, 2007.

PEREIRA, A.F.N., BARROS, I.C.L., SANTIAGO, A.C.P. & SILVA, I.A.A. **Floristics and geographical distribution of ferns and lycophytes of the Gurjaú Ecological Reserve, Pernambuco, Brazil.** Rodriguésia 62(1):1-10, 2011.

PIFANO, D.S., VALENTE, A.S.M., CASTRO, R.M., PIVARI, M.O.D., SALIMENA, F.R.G. & OLIVEIRA-FILHO, A.T. **Similarity between the habitats of the vegetation of Morro do Imperador, Juiz de Fora, Minas Gerais, based on the composition of its phanerogamic flora.** Rodriguésia 58(4): 885-904, 2007.

PINTO, L.P. & BRITO, M.C.W. **Dynamics of biodiversity loss in the Brazilian Atlantic Forest: an introduction.** In: Galindo-Leal, C. & Câmara, I.G. (Eds.). Atlantic Forest - Biodiversity, Threats and Perspectives. Pp. 27 - 30. SOS Mata Atlântica Foundation, São Paulo, 2005.

PRADO, J. **Pteridophytes of the state of São Paulo.** In: Bicudo, C. E. M. & Shepherd, G. J. (Eds.). Biodiversity of the State of São Paulo: Synthesis of Knowledge at the End of the 20th Century - Macroscopic Fungi and Plants. FAPESP, São Paulo. Pp. 49-61, 1998.

PRADO, J. **Pteridaceae** in Lista de Espécies da Flora do Brasil. Botanical Garden of Rio de Janeiro. Available at: <http://floradobrasil.jbrj.gov.br/jabot/floradobrasil/FB91978>. Accessed on: 16 May 2015.

PRADO, J.; RODRIGUES, C.D.N.; SALATINO, A. & SALATINO, M.F.L. **Phylogenetic relationships among Pteridaceae, including Brazilian species, inferred from rbcL sequences.** Taxon 56: 355-368, 2007.

PRADO, J. & SYLVESTRE, L. **Ferns and Lycophytes.** In: List of Species of the Flora of Brazil. Rio de Janeiro Botanical Garden. Available at: http://floradobrasil.jbrj.gov.br/jabot/floradobrasil/FB128483. Accessed on 24.03.2015.

PRADO, J; WINDISCH, P.G. **The genus Pteris L.(Pteridaceae) in Brazil.** Bol. Inst. Bot.(Sao Paulo), v. 13, p. 103-199, 2000.

Juiz de Fora City Hall. 2013. Lajinha Park. http://www.pjf.mg.gov.br/agenda_jf/parque_laginha/index.php Accessed 20.03.2013, PRICE, M.G. **Pecluma, a new tropical American fern genus.** American Fern Journal, p. 109116, 1983. PRYER, K.M., SCHUETTPELZ, E., WOLF, P.G., SCHNEIDER, H., SMITH, A.R. &

CRANFILL, R. **Phylogeny and evolution of ferns (Monilophytes) with a focus on the early leptosporangiate divergences.** American Journal of Botany 91: 1582- 1598, 2004.

RENZAGLIA, K.S., DUFF, R.J., NICKRENT, D.L. & GARBARY, D.J. **Vegetative and reproductive**

**innovations of early land plants: implications for a unified phylogeny.** Philosophical Transactions of the Royal Society of London B 355: 769-793, 2000.

ROLIM, L.B.; SALINO, A.. **Polypodiaceae Bercht & J. Presl (Polypodiopsida) in the Park Itacolomi State, MG, Brazil.** Lundiana, v. 9, n. 2, p. 83-106, 2008.

ROOS, M. **Mapping the world's pteridophyte diversity - systematics and floras.** In: Camus, J.M., Gibby, M. & Johns, R.J. (Eds.). Pteridology in Perspective. Royal Botanic Gardens, Kew. Pp. 29-42, 1996.

SALIMENA, F.R.G., CABRAL, A.F., FERNANDES, A.J., RODRIGUES, I.L. **Contribution to the floristic knowledge of Lajinha Park - Juiz de Fora - MG.** In: 56th National Botanical Congress, Curitiba. Abstracts of the 56th CNB, 2005.

SALINO, A. & ALMEIDA, T.E. **Pteridophytes from Jacupiranga State Park, SP, Brazil.** Acta Botanica Brasilica 22(4): 983-991, 2008.

SALINO, A., ALMEIDA, T. E., SMITH, A. R., GÓMEZ, A. N., KREIER, H. P., & SCHNEIDER, H. **A new species of Microgramma (Polypodiaceae) from Brazil and recircumscription of the genus based on phylogenetic evidence.** Systematic Botany, 33(4), 630-635, 2008.

SALINO, A. & ALMEIDA, T.E. **Pteridophytes.** In: Drummond, G.M., Martins, C.S., Greco, M.B. & Vieira, F. (Eds.). Biota Minas: Diagnosis of knowledge about biodiversity in the state of Minas Gerais - Subsidy for the Biota Minas programme. pp. 331-352. Biodiversitas Foundation, Belo Horizonte, 2009.

SALINO, A.; ALMEIDA, T.E. Ctenitis in Lista de Espécies da Flora do Brasil. Rio de Janeiro Botanical Garden. Available at: <http://floradobrasil.jbrj.gov.br/jabot/floradobrasil/FB115587>. Accessed on: 31 Mar. 2015

SALINO, A.; ALMEIDA, T.E. **Thelypteridaceae** in Lista de Espécies da Flora do Brasil. Jardim Rio de Janeiro Botanical Centre. Available at: <http://floradobrasil.jbrj.gov.br/jabot/floradobrasil/FB115525>. Accessed on: 07 Apr. 2015

SALINO, A.; JOLY, C.A. **Pteridophytes of three remnants of gallery forests in the JacaréPepira river basin, São Paulo State, Brazil.** Bulletin of the Ezechias Paulo Heringer Herbarium 8: 5-15, 2001.

SALINO, A., SEMIR, J. **Thelypteridaceae (Polypodiophyta) from São Paulo State:** Macrothelypteris **and** Thelypteris **subgenera** Cyclosorus **and** Steiropteris. Lundiana, v. 3, n. 1, p. 9-27, 2002.

SALINO, A., SEMIR, J. **Thelypteris subg. Meniscium (Thelypteridaceae-Pterophyta) in the state of São Paulo, Brazil.** Brazilian Journal of Botany, v. 27, n. 1, p. 103-114, 2004a.

SALINO, A., SEMIR, J. **Thelypteris subg. Amauropelta (Kunze) AR Sm.(ThelypteridaceaePterophyta) in the State of São Paulo, Brazil.** Lundiana, v. 5, n. 2, p. 83-112, 2004b.

SANTANA, D.J., FONSECA, E.M., NEVES, M.O. & CARVALHO, R.M.H. **A new species of Adelophryne (Anura: Eleutherodactylidae) from the Atalantic Forest, southeastern Brazil.** Salamandra 48(4): 187-192,

2012.

SANTIAGO, A.C.P. & BARROS, I.C.L. **Pteridoflora of the Charles Darwin Ecological Refuge (Igarassu, Pernambuco, Brazil).** Acta Botanica Brasilica 17(4): 597-604. 2003

SANTIAGO, A.C.P., BARROS, I.C.L. & SYLVESTRE, L.S. **Pteridophytes occurring in three forest fragments of an altitude marsh (Bonito, Pernambuco, Brazil).** Acta Botanica Brasilica 18(4): 781-792, 2004.

SANTOS, M. G., & SYLVESTRE, L. D. S. **Floristic and economic aspects of the pteridophytes of a rocky outcrop in the State of Rio de Janeiro, Brazil.** Acta Botanica Brasilica, 20(1), 115-124, 2006.

SCHNEIDER, H., PRYER, K.M., CRANFILL, R., SMITH, A.R. & WOLF, P.G. Evolution of vascular plant body plans: a phylogenetic perspective. Pp. 330-364. In: Developmental Genetics and Plant Evolution, Q.C.B. Cronk, R.M. Bateman, and J.A. Harris (eds.). Taylor and Francis, Philadelphia, 2002.

SCHNEIDER, H., RUSSEL, S.J., COX, C.J., BAKKER, F., HENDERSON, S., RUMSEY, F., BARRET, J., GIBBY, M., Vogel, J.C. **Chloroplast phylogeny of asplenioid ferns based on rbcL and trnL-F spacer sequences (Polypodiidae, Aspleniaceae) and its implications for biogeography.** Systematic Botany, v. 29, n. 2, p. 260-274, 2004.

SCHWARTSBURD, P.B. & LABIAK, P.H. **Pteridophytes from Vila Velha State Park, Ponta Grossa, Paraná, Brazil.** Hoehnea 34(2): 159-209, 2007.

Schwartsburd, P.B. Dennstaedtiaceae in Lista de Espécies da Flora do Brasil. Botanical Garden of Rio de Janeiro. Available at: <http://floradobrasil.jbrj.gov.br/jabot/floradobrasil/FB90936>. Accessed on: 13 May 2015 SMITH, A. R.. **Systematic of the neotropical species of Thelypteris section Cyclosorus.** University of California Publications in Botany, 59: 1-143, 1971.

SMITH, A.R., KREIER, H. P., HAUFLER, C. H., RANKER, T. A., & SCHNEIDER, H. **Serpocaulon (Polypodiaceae), a new genus segregated from Polypodium.** Taxon, v. 55, n. 4, p. 919-930B, 2006.

SOUZA, F.S. **Pteridophytes of Serra do Caparaó, Brazil: inventory and floristic relationships.** Dissertation (Master's in Ecology), Federal University of Juiz de Fora, 2012. Juiz de Fora

SOUZA, F.S., SALINO, A., VIANA, P.L. & SALIMENA, F.R.G. **Pteridophytes from Serra Negra, Minas Gerais, Brazil.** Acta Botanica Brasilica 26(2): 378-390, 2012.

THIERS, B. 2013. [Updated constantly]. **Index Herbariorum: A global directory of public herbaria and associated staff.** New York Botanical Garden's Virtual Herbarium. http://sweetgum.nybg.org/ih/. Accessed on 21.03.2013, 16h06min.

TUOMISTO, H. & POULSEN, A. D. **Influence of edaphic specialisation on pteridophyte distribution in neotropical rain forest.** Journal of Biogeography 23: 283-293, 1996.

TUOMISTO, H. & RUOKOLAINEN, K., KALLIOLA, R., LINNA, A., DANJOY, W., RODRIGUES, Z. **Dissecting Amazonian Biodiversity.** Science 269:63-66, 1995.

TRYON, R.M. & TRYON, A.F. **Ferns and allied plants with special reference to tropical America.** Springer-

Verlag, New York, 1982.

Tryon, R. M. & Stolze, R. G. 1993. Pteridophyta of Peru, Part V: 18. Aspleniaceae - 21. Polypodiaceae. Fieldiana Botany, new series 32: 1-190.

VASQUES, D. T.; PRADO, J. Campyloneurum C. Presl (Polypodiaceae) in the State of São Paulo, Brazil. Hoehnea, 38(2), 147-164, 2011.

XAVIER, S.R.S. & BARROS, I.C.L. **Pteridophytes occurring in fragments of Serrana Forest in the state of Pernambuco, Brazil.** Rodriguésia (54)83: 8-21, 2003.

WINDICH, P.G.; RAMOS, C.G.V.; 0LLGAARD, B. **Lycopodiaceae** in Lista de Espécies da Flora do Brasil. Rio de Janeiro Botanical Garden. Available at: <http://floradobrasil.jbrj.gov.br/jabot/floradobrasil/FB128549>. Accessed on: 31 Mar. 2015

WINDISCH, P.G.; SANTIAGO, A.C.P. **Cyatheaceae** in Lista de Espécies da Flora do Brasil. Rio de Janeiro Botanical Garden. Available at: <http://floradobrasil.jbrj.gov.br/jabot/floradobrasil/FB90854>. Accessed on: 31 Mar. 2015

WINTER, S.L.S.; SYLVESTRE, L.S.; PRADO, J. **The genus** Adiantum **(Pteridaceae) in the state of Rio de Janeiro, Brazil.** Rodriguésia, v. 62, p. 663-681, 2011.

**Annex**

Table 1. List of taxa occurring in Lajinha Park with habit and environment found, all collection numbers refer to collector L.V. Lima. Legend: Habitat: TE (Terrestrial); EP (Epiphytic); RU (Rupicolous); LI (Liana); HE (Hemi-epiphytic); Environment: IM (Interior of forest); BM (Edge of forest); AA (Anthropised area); BR (Brejo).

| Taxon | Habit | Environment | Material examined |
|---|---|---|---|
| Anaemiaceae | | | |
| A. phyllitidis (L.) Sw. | TE | BM/IM | 02 |
| A. hirsuta (L.) Sw. | RU/TE | BM/AA | 17 |
| A. villosa Humb. & Bonpl. ex Willd. | UK | /BM | 29 |
| Blechnaceae | | | |
| Blechnum austrobrasilianum de la Sota | TE | IM | 13,27 |
| B. acutum | RU/HE | IM | 57 |
| B. brasiliense Desv. | TE | BM | |
| B. cordatum (Desv.) Hieron. | TE | /BM/AA | 05 |
| B. occidentale L. | RU/TE | BM/AA | 09 |
| B. polypodioides Raddi | TE | BM/IM | 05,25 |
| Salpichlaena volubilis (Kaulf.) J.Sm. | TE/LI | IM | 58 |
| Cyatheaceae | | | |
| C. corcovadensis (Raddi) Domin | TE | BM | 53 |
| C. delgadii Sternb. | TE | IM | 01 |
| C. microdonta (Desv.) Domin | TE | IM | 54 |
| C. phalerata Mart. | TE | BM/IM | 60 |
| Dennstaedtiaceae | | | |
| Hypolepis repens (L.) C. Presl | TE | BM | 41 |

| | | | |
|---|---|---|---|
| Dryopteridaceae | | | |
| Ctenitis distans (Brack.) Ching | TE | BM/IM | 42 |
| Polybotrya speciosa Schott | HE | IM | 46 |
| Gleicheniaceae | | | |
| Dicranopteris flexuosa (Schrad.) Underw. | RU/TE | BM/AA | 34 |
| Gleichenella pectinata (Willd.) Ching | | | 45 |
| Sticherus lanuginosus (Fée) Nakai | TE | AA | 06 |
| S. nigropaleaceus (J.W. Sturm) J. Prado & Lellinger | TE | BM/AA | 20 |
| Lycopodiaceae | | | |
| Palhinhaea cernua (L.) Franco & Vasc. | TE | BM | 04 |
| *Phlegmariurus reflexus* (Lam.) 0llg. | TE | /AA | 08 |
| Pseudoycopodiella caroliniana (L.) Holub | RU/TE | IM/AA | 37 |
| Polypodiaceae | | | |
| Campyloneurum acrocarpon Fée | RU/EP/TE | IM | 49 |
| Microgramma squamulosa (Kaulf.) de la Sota | EP | IM | 48 |
| Pecluma paradiseae (Langsd. & Fisch.) M.G. Price | EP/TE | IM | 44,52 |
| Phlebodium pseudoaureum (Cav.) Lellinger | RU/EP/TE | BM/IM | 58 |
| Pleopeltis astrolepis (Liebm.) E. Fourn. | EP | IM | 35 |
| P. hirsutissima (Raddi) de la Sota | RU/EP/TE | BM/IM | 59 |
| Serpocaulon catharinae (Langsd. & Fisch.) A.R. Sm. | EP/TE | BM/IM | 43 |
| S. fraxinifolium (Jacq.) A.R.Sm. | EP/HE | IM | 50 |
| Pteridaceae | | | |
| Adiantum pentadactylon Langsd. & Fisch. | TE | IM | 16 |
| Adiantum raddianum C. Presl | TE | BM | 15 |
| Doryopteris nobilis (T. Moore) C. Chr. | RU/TE | BM/IM/AA | 22, 38 |
| Pityrogramma calomelanos (L.) Link | RU/TE | AA | 11 |
| P. schwakeana Christ | TE | BM/IM | 47 |
| P. vittata L. | UK | AA | 26 |
| Thelypteridaceae | | | |
| Macrothelypteris torresiana (Gaudich.) Ching | TE | BM | 30,31,40 |
| Thelypteris amambayensis (Christ) Ponce | TE | IM | 10, 28, 39 |
| Thelypteris dentata (Forssk.) E. P. St. John | TE | BM/IM/AA | 12, 32 |
| T. mexieae (C. Chr. ex Copel.) Ching | TE | IM | 51 |
| T. opposita (Vahl) Ching | TE | BR/AA | 07 |
| T. regnelliana (C. Chr.) Ponce | TE | BM/IM | 19,36 |
| T. rivularioides (Fée) Abbiatti | TE | BM/IM | 24 |
| Thelypteris serrata (Cav.) Alston | TE | BM/IM | 14 |

**Figure 3** - Some monilophyte species recorded in the study area: A) *Blechnum occidentale,* habit. B) *Anemia phyllitidis*, habit. C) *Campyloneurum acrocarpon*, habit. D) *Cyathea microdonta,* E) *Serpocaulon catharinae*, habit. F) *Pecluma robusta,* detail of abaxial face. habit.

MIX
Papier aus verantwortungsvollen Quellen
Paper from responsible sources
FSC® C105338

Printed by Books on Demand GmbH, Norderstedt / Germany